What they're saying about "The P

> *An empowering account of how citizen activists spoke up, demanded and created a healthy community. Susan Varlamoff provides a virtual guide for mothers who want to clean up their environments and more proof that together, we can make a difference.*
> **–Bella S. Abzug, co-chair,**
> **Women's Environment and Development Organization**

> *In addition to obvious regional appeal,* The Polluters *could serve as an instructive how-to manual for residents of communities confronting similar environmental hazards.*
> **–BOOKLIST, American Library Association**

> *. . . inspiring to the many front line fiahtᴇ⸱⸱*
> **–Lois Marie Gibbs, Execᴵ⸱⸱**
> **Citizen's Clearinghousᴇ**

> *This book is a must reₐ* ⸱⸱ᵘcial *decisions regarding our environ* ₋₉ ₘade in *a responsible manner without oul.* ⸱⸱ᵤᵢₑₑ.
> **–Mark Williams, Georgia Chapter of the Sierra Club**

> *. . . the telling of this intense journey lays out a blueprint of tenacious action for anyone who takes on any of the entrenched powers of the industrial/governmental complex of collusion.*
> **–Olin M. Ivey, Ph.D., Executive Director,**
> **Georgia Environmental Organization, Inc.**

> *The future environmental movement will find the lessons learned from her experience to be valuable for their future efforts..*
> **–Charles A. Cole, Professor of Engineering and Chair,**
> **Environmental Programs, Penn State University**

> *This is a compelling story superbly told by Susan Varlamoff. It tells how "the people" can arise, unite and win a victory of major proportions against a large corporation with all its money and connections.*
> **–Grant J. Merritt, Merritt, Furber & Timmer**

> *Susan Varlamoff creates a suspense story out of garbage. Susan spins out the all too true tale of a battle of titans. She brings to life a giant corporation and giant governments as they are confronted by a resolute woman..*
> **–Sidney Pauly, Representative, Minnesota State Legislature**

THE POLLUTERS

A COMMUNITY FIGHTS BACK

BY SUSAN JEZSIK VARLAMOFF

St. John's Publishing, Inc.

6824 Oaklawn Avenue ◆ Edina ◆ Minnesota ◆ USA ◆ 55435

On the cover
A backhoe digs up garbage while an Eden Prairie
firefighter hoses down the escaping fumes and gases from
an underground fire at Flying Cloud Landfill.
Seven fire departments fought the blaze
around the clock for seven days in January of 1980.

The cover photograph was taken by Don Church
and appeared in the St. Paul Pioneer Press
in January of 1980.

THE POLLUTERS
A St. John's Book/ Fall 1993
Copyright © 1993 by Susan Jezsik Varlamoff

Cover photo ©1980 St. Paul Pioneer Press.
Used with permission.

Published by
St. John's Publishing, Inc.,
6824 Oaklawn Avenue, Edina, Minnesota, 55435.

Printed on recycled paper

ISBN 0-938577-07-7 (paper bound)

PRINTED AND BOUND IN THE UNITED STATES OF AMERICA
First Edition.
0 9 8 7 6 5 4 3 2 1

363.73
V3/4p

101194

To the children of the earth,
born and unborn, that they may breathe clean air, drink
pure water, and eat the fruits of an untainted land.

The salvation of the world lies in the human heart, in the human power to reflect, in human meekness, and in human responsibility. We are still under the sway of destructive and vain belief that man is the pinnacle of creation and not just a part of it, and therefore, everything is permitted. We still don't know how to put morality ahead of politics, science, and economics. We are still incapable of understanding that the only genuine backbone of all our actions — if they are to be moral— is responsibility. Responsibility to something higher than my family, my country, my firm, my success. Responsibility to the order of Being, where, and only where, they will be properly judged.

VACLAV HAVEL

ACKNOWLEDGMENTS

John and Dorothy Jezsik, my parents, gave me the values by which I attempt to conduct the affairs of my life. They taught by their example that even during the darkest times, one must search out a path and press forward. When I was a child, they introduced me to nature on weekly Saturday morning walks in the woods. Also, many thanks are owed to my father who babysat his grandchildren in Pennsylvania so I could return to Minnesota and finish the work I began there.

Ivan, my dear husband, tolerated a crusading wife for many years and supported my obsession despite the time it took from our marriage. His ability to provide well for our family allowed me precious time to pursue the cause. Then, after it was over, he relived it once again, editing the first draft of my manuscript. Our children, Pierre, Neil, and Paul, too often had a mother attached to the phone and absent from home attending meetings. I gratefully acknowledge my family's forbearance during this time.

Throughout the course of the crusade, so many people provided me help and encouragement that it's difficult to name them all here. However, several people should be mention since their contributions were special.

Sidney Pauly worked ten years, first as a city councilwoman, then as a state representative, to stop the local landfill dump expansion and support our citizen's organization. As our case progressed, she initiated legislation that helped make Minnesota an environmental leader in the country.

Will Collette, field director for Citizen's Clearinghouse for Hazardous Waste, provided the winning strategy and emotional support I needed during our blackest hours of defeat.

Lois Gibbs of Love Canal fame, the role model we eco-activists all emulate, flew to Eden Prairie, Minnesota, and convinced people to continue the quest after my departure.

Drs. Dwain Warner, Joseph Mengel, and Doris Brooker generously donated their scientific expertise to our group. Many thanks, also, to Ed Crozier of the U.S. Fish and Wildlife Refuge, Chuck Moos of Upgrala Hunting Club, and Frank Ornstein of Clean Water Action for lending their help.

Greenpeace, Earth First, and Earth Protector Activists taught us fundamentals of planning and executing media grabbing demonstrations.

Citizen leaders, John Boyle, Stan Johannes, Cary Cooper, Peter Sadowski, John and Mary Kennedy, Louis Affais, Tim Homes, Scott and Mary Anderson, and Dick and Jerri Coller put in ungodly hours for years working to shut the dump.

Father Tim Power provided moral guidance and the back rooms of Pax Christi Catholic Church for our meetings.

City Council members Patricia Pidcock, Dick Anderson, Dr. Jean Harris, and Doug Tenpas pro-actively supported our citizen's group.

Governor Rudy Perpich, Attorney General Hubert H. Humphrey III, United States Senator Dave Durenberger, Senator Don Storm, Metropolitan Council Chair Sandra Gardebring, Metropolitan Council Representatives Dirk DeVries and Dottie Rietow, and Hennepin County Commissioner Mark Andrews, took a politically risky stance at our side.

To our many financial benefactors, my heartfelt thanks: Flagship, Naegele Outdoor Advertising Co., Residence Inn, Pax Christi Catholic Church, and Duane Pidcock provided much appreciated services.

To thousands of citizens who participated in this environmental quest, many of them unknown to me personally, I'm honored to have served with you.

I acknowledge the expert guidance provided by writer Rhonder Thomas Young, who believed the story was worth telling and gave me guidelines for writing, rewriting and editing my manuscript. And I appreciate the editing help provided by Sidney and Roger Pauly, Grant Merritt, Joe Mengel, Dick and Jerri Coller, my husband, and my father.

Finally, my publisher and editor, Donna Montgomery, agreed to launch this book so others may take hope in their own ability to make a small difference in the world too. Thank You!

CONTENTS

FOREWORD

In 1988, my longtime lawyer friend Grant Merritt told me he was retained by the City of Eden Prairie as a special legal consultant. Earlier, Grant worked with an Eden Prairie citizen's group headed by Susan Varlamoff, opposing expansion of the Flying Cloud Landfill. Susan and her group of activists were trying to stop a billion dollar garbage company from expanding a leaking, toxic landfill close to their homes.

The details of Grant's new project caught my interest. We had collaborated before on other environmental projects; Grant did legal work, and I supplied scientific information, my profession being geology and environmental consultation.

Without any prodding, I rolled out my maps and reviewed topography around the Flying Cloud landfill. How extremely odd, I thought, that this landfill was sited in highly permeable glacial outwash. Moreover, it's at the crest of an eroding bluff, sits over a major drinking water aquifer, and is adjacent to a wildlife refuge and a major regional airport. Incredibly, a residential community developed around it. How could such a toxic, leaking, poorly located landfill be considered a candidate for expansion? It seemed totally absurd.

Our earth is a dynamic system with matter constantly in motion. Modern engineering technology can only temporarily overcome nature. Even if a garbage company carefully engineered a landfill or its expansion, the work would eventually fail, due to natural decay processes and land shifting.

If the landfill expansion case had been won, a dangerous precedent would have been set to allow continuation of a hazardous operation in a

11

residential neighborhood.

I agreed to work with Grant, Susan, and the citizen's group. I was especially impressed at how Susan electrified and motivated her large following, mostly mothers of Eden Prairie, to protect their families from a seriously threatening situation.

In this strenuous struggle, one can see incarnated the spiritual dimension of humankind, how it's possible to construct a cutting edge of effective action to improve circumstances of one's life through careful questioning, analysis, enduring determination, and a shared vision of a better life for all.

Susan Varlamoff is a pioneer of grass roots action to rectify serious environmental abuses. It's amazing how this full-time homemaker and mother of three came from anonymity to electrify a community of some 40,000 people. However, It's going to take a good many more Susan Varlamoffs to save us.

Joseph Mengel, Jr., Ph.D.
Geologist and Environment Consultant

PREFACE

Through this book, you'll experience the human drama of ordinary people fighting to shut down a leaking, toxic dump threatening their Midwestern neighborhood. You'll follow them into their living rooms for secret strategy sessions, go before regulatory agencies, hike through hallowed halls of our state capitol, wait behind closed door sessions in court, then move on to the highest levels of government in Washington D.C. You'll feel their rage, their pain, their defeat, and their triumphs as they battle the Minnesota state government and the world's second largest garbage company.

These homemakers, teachers, farmers, construction workers, priests, ministers, business people, mothers, and fathers are all individuals with whom you can identify because they live in nearly every community and share a universal desire to raise their children in safe surroundings and protect their homes. When these citizens realize Big Business interests could crash their dreams, they fight back. However, ordinary people lack honed skills of professional lobbyists, financial resources of large corporations, or the cunning of politicians and the technical knowledge of scientists. It's an extraordinary desire to protect their families that drives them to succeed.

This book contains detailed plans of one community's fight. The formula is simple and carefully laid out. You'll experience staging a protest, engaging various media, raising money, setting up a non-profit corporation, hiring a lawyer, and dealing with politicians. Anyone interested in duplicating this effort will have fundamentals with which to begin.

Lastly, the many struggles of these ordinary men, women, and children will inspire you as they throw caution to the wind and strive together for the common good to preserve their beautiful prairie land named after The Garden of Eden. *The Polluters* doesn't pretend to be an intellectual treatise exploring scientific solutions for saving the earth's ecosystem from impending collapse. Instead, this story is meant first to inform you, secondly to prompt you to take a stand in your own community against environmental pollution. Citizen collective action may be the only hope we have of saving our civilization.

THE
LAY
OF THE
LAND

The Flying Cloud Landfill site in Eden Prairie, Minnesota

1945

Flying Cloud Drive (Hwy. 169)

Glacial depression

Minnesota River Bluffs

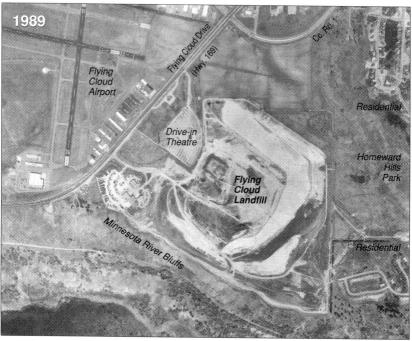

1989

Flying Cloud Drive (Hwy. 169)

Co. Rd. 1

Flying Cloud Airport

Drive-in Theatre

Flying Cloud Landfill

Residential

Homeward Hills Park

Residential

Minnesota River Bluffs

AERIAL PHOTOGRAPHS BY MARKHURD. USED WITH PERMISSION.

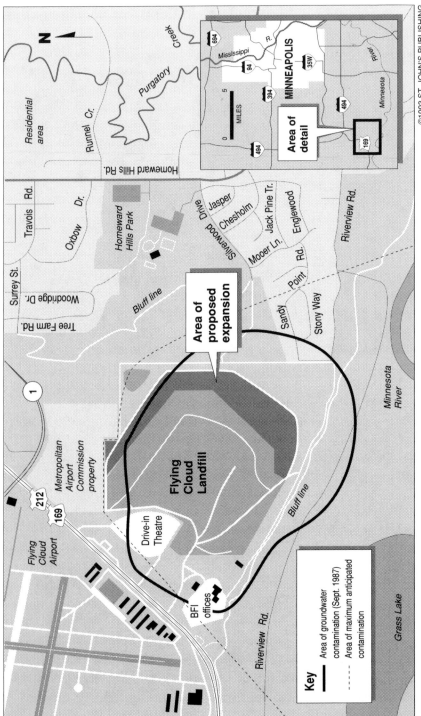

N

Residential area

Purgatory Creek

Runnel Cr.

Homeward Hills Rd.

Travois Rd.

Oxbow Dr.

Surrey St.

Tree Farm Rd.

Woodridge Dr.

Homeward Hills Park

Silverwood Drive

Jasper

Chesholm

Jack Pine Tr.

Englewood

Mooer Ln.

Point Rd.

Sandy

Stony Way

Riverview Rd.

Bluff line

Area of proposed expansion

1

Metropolitan Airport Commission property

212

169

Flying Cloud Airport

Drive-in Theatre

Flying Cloud Landfill

Bluff line

Minnesota River

BFI offices

Riverview Rd.

Grass Lake

Key

Area of groundwater contamination (Sept. 1987)

Area of maximum anticipated contamination

Area of detail

694

Mississippi R.

94

MINNEAPOLIS

394

35W

494

Minnesota River

169

494

MILES

0   5

©1993 ST. JOHN'S PUBLISHING

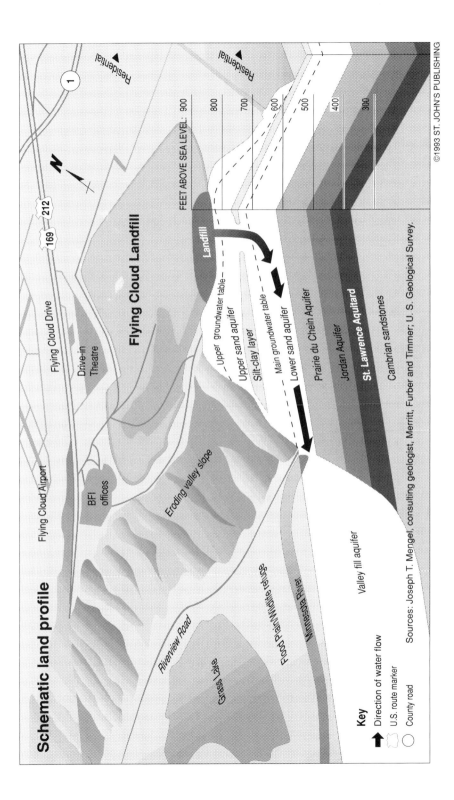

Schematic land profile

Flying Cloud Landfill

Residential

Residential

1

169 212

N

Flying Cloud Drive

Drive-in Theatre

Flying Cloud Airport

BFI offices

Eroding valley slope

Riverview Road

Flood Plain/Wildlife Refuge

Minnesota River

Grass Lake

Valley fill aquifer

Landfill

FEET ABOVE SEA LEVEL: 900

800

700

600

500

400

300

Upper groundwater table
Upper sand aquifer
Silt-clay layer
Main groundwater table
Lower sand aquifer
Prairie du Chein Aquifer
Jordan Aquifer
St. Lawrence Aquitard
Cambrian sandstones

Key

➤ Direction of water flow

◯ U.S. route marker

◯ County road

Sources: Joseph T. Mengel, consulting geologist, Merritt, Furber and Timmer; U. S. Geological Survey.

1 | OMINOUS LETTER

van and I kissed our three young sons good night and turned them over to a sitter for the evening. We bundled up against the cold Minnesota February of 1982 and joined a long procession of cars winding its way up Homeward Hills Road en route to a neighborhood meeting. A letter had arrived from a Minneapolis law firm that week stating its client, Woodlake Sanitary Services, Inc., (WSSI) wanted to expand its operation at the Flying Cloud Sanitary Landfill. WSSI wished to discuss the proposal with nearby residents. We were stunned! An expansion to *what* landfill?

Ivan and I had no clue a garbage dump existed two blocks from our home, adjacent to our recently built subdivision in the city of Eden Prairie. It never occurred to us to question our realtor or anyone else about the permanence of our community's breathtaking, rolling prairie stretching to the horizon. And we never thought to research the location of the nearest garbage dump. Like many young, first time homeowners in the area, we naively trusted the glorious vision before us.

Until now, our existence had been a simple one. As thirty-year old parents of Pierre, Neil, and Paul, we focused our energies on paying our monthly mortgage, vaccinating our boys, toilet training Neil, changing Paul's diapers, and accomplishing the myriad of other duties required to care for a five-year old, a three-year old and a six-month old. There remained little extra time and energy to worry about matters outside our home and family. However, the existence of a landfill and its possible expansion suddenly exploded on us, forcing us to face a potentially grave situation nearly at our doorstep. We had no choice but to look beyond our fence.

Telephone lines burned in the Bluff's West development of about fifty households as neighbor called neighbor to discuss the ominous Woodlake letter.

"Can you believe there's a garbage dump next to our development?"

"No, and I'm mad our realtor never told us about it. We can't afford to move again. We just bought this home."

"This will kill our property values."

"What are we going to do?"

Equally distressed were residents of Hillsborough, a subdivision located directly across a park from us, and on the northern side of a hill camouflaging the offensive landfill. Outrage and worry consumed our conversations, too, as we discussed the proximity of an expanded garbage dump.

We had no idea how this situation would impact our lives. Nevertheless, we spent hours thinking about frightening possibilities. Recent headline news that President Carter provided funds to relocate families living next to Love Canal, a hazardous waste dump causing sickness and death to nearby residents, sparked our fertile imaginations.

"DEATH FROM DUMP STRIKES LITTLE HOUSE ON THE PRAIRIE" might scream out from page 1 of the Minneapolis *Star Tribune*, we mused. Were we overreacting? Could this landfill, in fact, pose a risk to our homes and families? Was there more than just household waste in this garbage dump? We would not know the answer to these and many other questions for years. However, we all agreed to send at least one member from each household to an upcoming meeting to find out what we could.

To represent the Varlamoff family, both Ivan, an engineer, and I, a trained biologist, decided to attend and learn about the Woodlake proposal using our respective disciplines to help us comprehend the environmental impact. As we drove to Hennepin County Technical Center, the meeting location, I questioned my husband about the meeting's agenda. "What do you think will happen?" I asked Ivan.

"We need to wait and hear what this company expects to do," Ivan replied. "Since we have no information yet, let's not try to second guess the outcome."

How sensible, I thought. But then again, Ivan was always the quiet, cool, rational partner in our marriage, often described by colleagues as having his feet planted firmly on the ground. In contrast, I was a light-headed optimist with my eyes on the stars. I often reacted emotionally to each new crisis, had strong opinions on most issues, and could willingly debate them for hours. We were a good team. Ivan pulled me down out of the stratosphere, and I encouraged him to take flights of fancy.

After arriving at our destination, Ivan parked our blue Chevrolet station wagon near our neighbors' cars and we walked into the school with friends, exchanging views about the proposed landfill expansion. Once inside, we followed signs directing us to an auditorium, a room large enough to hold at least seven hundred people. The floor descended steeply to a raised stage where several men were busily setting up a media presentation for the evening's program. Ivan and I took seats somewhere in the middle rows and quietly waited for the show to begin.

After some 150 people had filed in, a sandy haired, well dressed, good looking man about thirty-five years old asked for the audience's attention.

"Good evening. My name is Dick, and I am a lawyer with the firm of Larkin and Hoffman," he announced, smiling broadly. "Tonight my colleague, Roy, an engineer with a Chicago firm, and I will present Woodlake Sanitary Service's landfill expansion plans."

Dick went on to explain that Woodlake's 100 acre landfill had reached near capacity and the company wanted to continue operations for ten more years. To accomplish this, they applied to the government for a permit to enlarge their facility. He then turned over the microphone to Dr. Roy to describe the technical aspects of the proposed project.

In a quiet, monotone voice, Dr. Roy proclaimed Woodlake's sanitary landfill expansion to be state-of-the-art and showed us various specifications on his multicolored flip charts and overhead transparencies. He pointed out technical features of this marvelous megadump and presented data to validate his assumptions. Dr. Roy next informed us that this expansion would be built on top of and adjacent to the existing landfill. Members of the audience exchanged skeptical looks.

We began feeling this whole demonstration was an elaborate effort to confuse us with scientific jargon. Yet several unsettling facts surfaced through the murky performance, like oil in water. Suspicion hung in the air and a low rumble broke the silence.

"Does this company think we're stupid?" I whispered to Ivan. "Do they really think we'll accept an expansion under these conditions."

Probably sensing the crowd's displeasure, Dick took back the microphone from Roy and cheerfully gave us "good news:" Woodlake's end use plan. For our recreational pleasure, the company intended to build an 18-hole golf course on top of the entire 240 acre filled landfill site. As Dick expounded on the merits of this "end use plan," and the prospect of perhaps even signing up citizens for tee times, the audience came unglued. Hands flew up in the air. One man angrily pointed out that the expansion would move the landfill closer to our homes and double its size. Undaunted, smiling, and running his fingers through his hair, Dick once again tried to divert the group's attention back to the golf course plan. The citizens, however, persisted with their line of questioning, and the scene progressed from mildly adversarial to overtly hostile. A shouting match ensued.

"Are you an expert?" shrieked one woman to Dr. Roy.

"Yes," he quietly answered.

"What about methane gas migrating off the landfill into our homes?" she wailed, but Dr. Roy insisted he had addressed all our concerns in the plans we just saw. There was no need for alarm.

People remained unconvinced.

"This facility will draw children to it like a magnet," stated Cary Cooper, a young father. "With so many in our neighborhood, I cannot believe you would consider a landfill expansion this close to our homes."

Dr. Roy reasserted his view that this expansion would be safe for children, but there were still no converts to the plan. Questioning continued until Dr. Pete Sadowski, our neighborhood biochemist, stood up and let loose his Polish temper. Wagging his finger for emphasis, Pete strongly challenged Dr. Roy's data. He was probably one of the few people in the room who could see past the colors and curves on Woodlake's slick charts and realize this proposed expansion could potentially harm us. Unperturbed by Pete's verbal attack, Dr. Roy defended his scientific models.

This raucous debate continued as more and more people yelled their discontentment with Woodlake's proposal. Despite Dick's and Dr. Roy's gallant efforts to promote the landfill expansion with its golf course end use plan incentive, residents stubbornly refused to accept it.

Ivan and I sat in stunned silence, overwhelmed by this theater of the absurd. We asked no questions and made no comments. Recent events had moved faster than our ability to absorb them. We had barely finished unpacking boxes after recently moving from Phoenix, when we discovered a landfill expansion could crush our dreams for a peaceful, healthy existence on the prairie. Having a golf course within walking distance of our home excited us, but then, what did we know about the ramifications of living near a garbage dump? It just didn't feel right, so we refrained from buying our golf clubs just yet.

Dick's polite smile seemed like a sham. How could he stand there and grin as citizens hurled angry remarks at him? I suspect even Jesus Christ did not smile when a mob of executers went after him. And if Dick was convinced this state-of-the-art landfill expansion was safe for the neighborhood, then why not debate its merits? Why push the golf course proposal down our throats? What was this company trying to hide?

Dr. Roy did little else to inspire confidence. He appeared to operate on another level: one of complex formulae and engineering models. How would his projected plans stand up against the elements of nature over time? And what about methane gas migrating off the landfill? Could it not explode in our homes?

We were enraged to think that Woodlake would take for granted our acceptance of their ten-year landfill expansion without reservations. This golf course end use plan seemed like a bribe for gaining acceptance. The whole deal smelled bad; like putrid, rotting garbage.

After a verbal battle of several hours, John Boyle, a neighbor who lived a trash bag's throw from the dump, defiantly stood up. He faced Dick and Dr. Roy and declared, "Gentlemen, this expansion will not be as simple as you anticipate. I intend to fight it."

With these remarks, the meeting adjourned and we hastily put on our coats, jackets, hats and gloves, and poured out into the cold night, disturbed by the evening's events. Something felt terribly amiss. We needed to find out what it was.

2 | LITTLE HOUSE ON THE PRAIRIE THREATENED

"Surely the Garden of Eden couldn't be more beautiful," exclaimed Elizabeth Fry Ellet, when she saw the land in 1853. This young author visited Minnesota at the request of M.Y. Beach, editor-in-chief of the *New York Times* and a well-known person in the literary world. He asked her to research the area for a book, so she traveled up the Minnesota River to look around with another distinguished author of the time. At one point along the journey, they climbed the river bluffs to the upper prairie. When Mrs. Ellet set eyes on the virgin prairie in full bloom, she was enchanted by the beautiful flowers. She described the area as the "garden spot of the territory," and on returning to St. Paul, requested that officials name the land "Eden Prairie."

Today, on this former paradise, stands the Flying Cloud Landfill, a leaking, toxic garbage dump surrounded by barbed wire. The virgin prairie now stands in ruin. Refuse suffocated flowers in 1970, when the landfill took in its first truckload of garbage. This 240 acre site, measuring 2,080 football fields in surface area and one football field deep, ranks as one of largest garbage dumps in Minnesota. To appreciate its magnitude, one must peer down at the landfill from a helicopter flying overhead.

The county chose Eden Prairie for its garbage dump because it contained a natural "kettle" left thousands of years ago by melting glaciers. This 80 acre depression, 90 feet deep, was created by the melting of a detached block of glacial ice, which with its surrounding silt and clay layer was submerged by the outwashed sands of the land. About 12,000 years ago, the meltwaters released three layers of granular sediments 200 feet thick. The top layer consists of unconsolidated sand, underlined by an irregularly thick layer of consolidated sand and below which is an irregularly thick layer of clayey silt. These granular subsoils act as a sieve by allowing water to quickly dissipate through to the drinking water aquifers below. Unlike many of these glacial depressions which later became Minnesota's 10,000 lakes, this one never held water.

Groundwater underneath the landfill flows slowly south into the Minnesota River and an adjacent wildlife refuge. This immense refuge contains rare species of plants and animals including our national bird, the bald eagle. Its fragile ecosystem requires a constant supply of fresh water.

Flying Cloud Landfill is perched on the crest of an eroding bluff which overlooks this spectacular landscape. Over time, natural forces have slowly washed the landfill slope into the river valley below. Sometimes the process is slow and inconspicuous, and at other times massive chunks of land crash down.

In the pre-dump era, after the glacial ice melted and before white settlers arrived on the scene, this land was the undisputed homeland and hunting ground of the Lakota. These indigenous people traveled the waters by canoe and lived off the animals and plants of the forest. The mighty Lakota nation was divided into bands, and the band known as the Mdewakanton Sioux wandered south and ultimately settled along the Minnesota River in the area of Eden Prairie.

The Mdewakanton Sioux lived well on the land of lakes and streams. Using bone hooks and spears they caught fish and turtles. Woods abounded in wild game. Flocks of wild geese, ducks, and other aquatic birds settled on the lakes, while muskrat, beaver, and mink lived along the shores.

People ate wild berries, plums, and roots in season and stored some for winter. Because cranberries grew abundantly in lowlands around the lakes, they were gathered in the fall. Wild rice was also plentiful in valley lakes along the Minnesota River. To the Mdewakanton Sioux this land was "Wa-se-cha," the land of plenty.

According to Jonathan Carver, an early explorer, these people traditionally placed their dead upon stagings overlooking lakes, rivers or other beautiful scenery which they enjoyed while they were living. They would leave bodies there for a certain time, then bury the parched bones in mounds with other family members. The landfill bluff contained several burial mounds that survived the farmer's plow — but not the hazards of a garbage dump.

Eventually the Mdewakanton Sioux were forced to sign away their land to white settlers in the treaties of Traverse des Sioux and Mendota in 1851. Covered wagon trains then came rumbling across the prairie with families eager to farm the land. Early pioneers planted grain, red clover, apples, and raspberries in the rich, black soil. They grazed large herds of cattle on the hillsides. For more than a hundred years, Eden Prairie remained a farming community governed by a township board.

Harsh weather consisting of cold, snowy winters, and tornadoes and droughts in spring and summer forced a spirit of camaraderie to develop among the people. Farmers shared their workloads and their lives with each other as they battled the elements to maintain their livelihoods. As a result, a strong community spirit took hold and still exists today in Eden Prairie.

Laura Ingalls Wilder chronicled the life of the early Minnesota farm family in her famous book, *Little House on the Prairie*, which later became a weekly television program. In fact, Laura lived in Walnut Grove during the 1870's, not more than 100 miles southwest of Eden Prairie. The grassy hill that Laura's television character runs down during the opening credits resembles the rolling prairie next to the landfill's Bluff's West subdivision.

As many farmers became unable to economically maintain their way of life, they began signing away their heritage to developers. Since Eden Prairie is located just a short, fifteen minute car ride from downtown Minneapolis, the area became prime real estate for business people commuting daily to and from work in the Twin Cities. As the population exploded, so did household waste. After the developers came Waste Management, Inc., a newly formed garbage company. In 1970, the company bought farmland along the Minnesota River bluffs from a widow to serve as a site for a new sanitary landfill.

Before the advent of landfills, people burned their household garbage outside in the open air. Industrial waste was indiscriminately dumped on the land and in lakes, streams and the ocean. But when the Federal government began to realize that such practices were adversely impacting public health and safety, they decided to regulate industry.

In 1970, the United States established the Federal Environmental Protection Agency to study pollution and write laws to reduce the release of toxic compounds into the air, water, and soil. As a result, engineers of the day designed the sanitary landfill to achieve safer disposal of household and hazardous waste. This became the recommended method for handling garbage.

Each state was then required to create their own agency to enforce federal laws. Minnesota formed the Pollution Control Agency (MPCA) and a young, aggressive lawyer named Grant Merritt became one of its first directors. He fearlessly prosecuted big business and forced compliance with regulations. Under his administration, citizens in northern Minnesota won a precedent-setting environmental case against Reserve Mining Company. They stopped the company from dumping taconite tailings into Lake Superior. Thirteen years later, Eden Prairie citizens hired Grant to prevent the world's second largest garbage company from expanding its garbage dump.

In 1970, burying garbage in a sanitary landfill simply meant dumping it into an unlined, bulldozed hole in the ground, then covering it over with a layer of soil. Scientists of this era considered such a primitive disposal method as state-of-the-art. High tech, engineered landfills with a lining belonged to a future 15 years away.

In time, engineers and scientists came to realize how little they understood about hydrology, the study of groundwater dispersion through the soils. Eventually, unmonitored, unlined landfills leaked toxins into groundwater. But how did toxins find their way into landfills in the first place? Some of these toxins came from paint cans, nail polish remover and other household waste contained in ordinary garbage, but the majority arrived illegally. Mixed together in a landfill, rainfall washed all these contaminants into underlying aquifers in dangerous concentrations. Thus, "sanitary" became an embarrassing description for the landfills' resulting pollution.

This out-of-sight, out-of-mind garbage disposal method eventually proved disastrous. Some scientists estimate that leaking sanitary landfills contaminate up to 70% of the country's groundwater, our principal drinking water source. Perhaps, "state-of-the-dinosaur" more accurately defines these primitive dumps.

In February, 1970, the Eden Prairie Planning Commission reviewed Waste Management's landfill plan and ordered studies to determine if the community's site was suitable for garbage disposal. A report prepared by the Chicago engineering firm, Greeley and Hanson, calculated the volume of the kettle at 3100 acre feet and figured that with an average daily dumping of 4,000 cubic yards of garbage, five days a week, "the total life of the site is estimated to be about 12 years."

The report added that the leachate (contaminated water produced by the rotting garbage) "will be subject to dilution and purification in the soil. It is expected that any contaminants in the groundwater will be reduced to below detectable limits by the time they reach the river." .... "Field examinations of sanitary landfills have shown that pollutants from the fills are not detectable after the fluid has moved 500 feet in fine sand."

D. K. Ripple, a landscape architect group from Tonka Bay, Minnesota, prepared an end use plan. They estimated the life of the landfill at 8 to 10 years. Both Greeley and Hanson and D. K. Ripple concluded that the end date for garbage dumping in the Flying Cloud Landfill would be 1982.

Eden Prairie planning commissioners felt satisfied with these reports and recommended the installation of a sanitary landfill in their community on April 7, 1970. Town fathers would live to regret this day, a day of infamy.

A quote from the Sun Newspaper described the situation this way: "Hennepin County has the need and Eden Prairie has the hole according to Waste Management, Inc., of Chicago. The site will fulfill approximately 1/3 of the county's needs for 8-10 years."

Following the city's recommendation, the Metropolitan Council, a county planning agency, held a public hearing to take citizen's comments. Dr. Dwain Warner, a noted ornithologist, who painstakingly catalogued plant and animal species in the river valley, warned officials that this site was unsuitable for a landfill due to its location on an eroding bluff and its proximity to a wildlife refuge. He predicted, "This landfill will be costly and dangerous to the people of Minnesota."

Dr. Warner stated that the landfill would probably contaminate groundwater flowing into wetlands. Polluted water would then be drunk by fish and animals in the refuge and adversely affect the entire food chain there. He further stated that the problem could take on international proportions. Explaining how many migratory birds use the refuge for a short stay on flights to Canada, South America, and so on, Dr. Warner concluded that the Flying Cloud Landfill could have the notorious distinction of infecting fauna and flora the world over.

These astute objections to the proposed landfill site did nothing to influence the outcome of the government's decision. Metropolitan Council officials voted "yes" to garbage dumping in Eden Prairie's "kettle". They said it conformed with the Solid Waste Disposal section of the Metropolitan Development Guide. Deeply disappointed, Dr. Warner dreaded he might one day witness a silent spring in the Minnesota River Valley, an area he profoundly knew and loved. Author Rachel Carson's words might yet prove to be prophetic: "To still the song of the birds and the leaping of the fish in the streams."

On August 27, 1970, Floyd J. Forsberg, Director of Solid Waste for the Minnesota Pollution Control Agency, signed the permit granting Waste Management Inc. permission to operate a municipal sanitary landfill in the Village of Eden Prairie. Under "General Conditions," the permit stated that "No major alterations or additions to the disposal system will be made without the written consent of the Agency."

Opening day at Flying Cloud Sanitary Landfill was a time of jubilation for Eden Prairie citizens. It began with an early morning breakfast celebration attended by key business and political leaders, followed by an afternoon extravaganza. School children were bussed to the site to watch Miss America cut the ribbon while the high school band played a command performance. All the town's movers and shakers showed up for this historic moment as T.V. cameras recorded the euphoria.

Little Maureen Pauly, dressed in her lime green pant dress, excitedly rushed home after school to tell her Mom about the goings on. "I think I'm on the evening news with my class at the landfill," she exclaimed breathlessly. Then somewhat perplexed, she went on to relate how she saw a new, environmentally safe way to dispose of garbage: a giant hole in the ground. Maureen added that layers of trash and dirt would fill the space. On hearing the story, her Mom, a natural science major in college, expressed misgivings about the new facility.

What a shining moment for this village of two thousand people! The new enterprise promised increased revenue for citizens, and an opportunity to be on the cutting edge of a state-of-the-art technology: the "sanitary landfill."

Because business at the dump in those early days wasn't terribly brisk, it closed for several months. Apparently, county garbage was being diverted to an unlicensed dump. Then, on March 13, 1972, Waste Management, Inc. sold their landfill to International Disposal Corporation (later changed to Woodlake Sanitary Service, Inc.), a wholly owned subsidiary of Browning Ferris Industries (BFI). In making the purchase, BFI accepted all land use and dump operation commitments made to the village by Waste Management, Inc. (Please note that the names Woodlake Sanitary Service, Inc. (WSSI) and Browning Ferris Industries (BFI) are used interchangeably, since activities at the Flying Cloud landfill were often directed by the parent company, BFI.)

As the population continued to grow, so did its garbage production. Now trucks drove through the Flying Cloud Landfill gate in a steady stream, emptying their contents into the great pit. No one thought to check on activities at the local dump, so all was quiet in Eden Prairie during the early 1970's.

Directly across the highway from the Flying Cloud Landfill, is Flying Cloud Airport, which serves small aircraft. In 1974, the Federal Aviation Administration (FAA) sent a letter to the landfill operators informing them of hazards caused by birds flying off the landfill into airplane windshields and propellers. The FAA explained that a regulation prohibited landfills from being situated closer than 5,000 feet to a runway used by piston type aircraft, and Flying Cloud Landfill violated this law. No problem. BFI lobbyists obtained the necessary variance to exclude their facility from the regulation.

Then in June of 1978, nature struck a major blow to the landfill bluff.

A chunk of land accompanied by garbage fell into the river valley. The Peterson family, other farmers in the area, and Dr. Dwain Warner, a wildlife scientist, were among the few to notice this change in landscape.

By the end of the 1970's, the local garbage dump no longer represented a quiet business affair. The town's fire fighters were kept busy extinguishing many fires that burned there. Between 1979 and 1980, four big blazes occurred at the landfill. One epic inferno lasted seven days and burned garbage 100 feet deep. Walter James, a local firefighter described the "bright yellow, white, blue, and green flames, . . nothing like a normal fire." He said he and his colleagues saw red bags of medical waste and stepped on used syringes while fighting the fire. They also observed "several hundred 55 gallon drums standing in black liquid."

Another firefighter vividly recounted the spectacle for me as he saw it ten years earlier from a helicopter window. He said giant flames leapt out of a huge black pit in the ground, lighting up the dark night. Then smoke billowed to the sky choking the air.

Eyewitness accounts of medical waste and chemical drums at the landfill site provided townspeople with their first concrete evidence that operations at the landfill did not jive with the permit. Obviously, illegal dumping activities had occurred. This information, however, never reached the eyes and ears of Eden Prairie City officials and the public, because years later, no one seemed aware of the abuse. Like the toxic waste in the dump, this knowledge remained buried.

In 1979, a representative from BFI's Woodlake Sanitary Service appeared at an Eden Prairie City Council meeting to announce closure of the landfill in approximately three-and-one-half years. Sidney Pauly, who was a council member at the time, told me the council believed this promise and voted to approve development of single family homes next to the landfill.

Then, in what seemed to be a schizophrenic move, BFI lawyers next notified the Pollution Control Agency in St. Paul that they planned to expand their nearly full landfill.

Moving to Jack Pine Trail

In March, 1982, Ivan traveled alone to Minnesota to purchase a home for our family. His company was transferring us to the area. I stayed behind in Phoenix to care for our three young children, one of whom was a nursing infant. Before Ivan's trip, we discussed the kind of house we wanted, so I trusted his judgement.

Ivan found just the right model, a new four bedroom colonial, halfway down a gently rolling hill, and within walking distance of a park. It had a cozy family room off the kitchen, complete with a fireplace to warm us on those legendary, cold, blustery Minnesota nights I'd read about. After his return from Minnesota, Ivan described for us in detail the Eden Prairie countryside and our house on Jack Pine Trail.

"The prairie extends as far as the eye can see. There are so many open spaces where the boys can play, and so many paths where Pierre and Neil can ride their new bicycles. Unfortunately, there are few trees for climbing. But the boys will enjoy running down the grassy hills in the summer and sledding down them in the winter."

Ivan snapped Polaroid pictures inside and outside the house to give the boys and me the opportunity to see our future surroundings. He even brought back samples of the chocolate brown carpet, tan linoleum, and off-white counter tops. We loved our little white house on the prairie even before we stepped through the doorway. It seemed the ideal place to raise a family. I imagined watching my boys pass many happy hours frolicking on the hills as Laura Ingalls had done a century earlier.

In July of 1982, after the moving vans left 12209 Jack Pine Trail, we quickly unpacked our boxes and sodded the yard. My mother-in-law, who flew in from Belgium to help us, bought each of us a sapling since our property had no trees. So we began our life in Eden Prairie nurturing children and trees on our land.

Neighbors visited often that summer and we chatted for hours outside, watching our children play until sunset. Jerry Wagner and myself, both proud mothers of six-month old infants, often brought our babies outside with us while we sat on my front lawn to talk. In what seemed a joint conspiracy, these babbling adventurers would often scramble toward the road, requiring Jerry and me to scoop them up every few minutes to prevent their being squashed by a car.

Sometimes, as a family, we set out to explore the neighborhood. This required getting out our red wagon, a gift from Great Grandpa, and loading it down with our boys and their toy trucks. Up and down hills I pulled the wagonload of children traveling from Jack Pine Trail to

Winter Place and back again, a distance of about three blocks. We noticed the neighborhood was only partially developed; many empty lots still remained. Except for the laughter of a few children, the area was quiet. As Jerry liked to say, "So quiet you can hear a bird fart."

What I enjoyed most that first summer, however, were the sunsets. Since the landscape was mostly flat with few trees, the horizon was open. The sky was often clear and bright blue during the day because little rain fell during June, July, and August. Such weather is a farmer's curse. But for me, a sunset lover, it was a dream.

By evening, the sky became awash in soft pinks, purples, reds, and oranges. Swirled together in the fading light, they created a massive impressionist painting. The shades of colors and patterns changed each day to produce a new masterpiece. Lingering outside on warm evenings was ample excuse to exchange small talk with neighbors. It was a sensual experience.

The first wave of activism

Just six months after settling into our new life on the prairie, we received the infamous 1982 letter from the Larkin, Hoffman law firm inviting us to attend a meeting to discuss Woodlake's landfill expansion proposal. This meeting quickly became "the great dump debate."

Eight to ten days after this showdown with BFI's lawyer Dick and Dr. Roy, John Boyle, Cary Cooper, Stan Johannes, and Pete Sadowski sat down together to form the Eden Prairie Citizen's Coalition (EPCC). John became president, Stan, vice-president, Cary, treasurer, and Pete, secretary and scientific expert. Their talents included accounting, banking, and biochemistry. They set several goals. To monitor the permit process for the landfill expansion, they decided one of them must be present at every meeting involving the issue. Realizing it would be a lengthy process and an enormous task, they believed delay tactics would force Woodlake to get fed up and give up its expansion plans. So the EPCC set the task of looking for ways to delay Woodlake.

Pete, the biochemist, studied various Woodlake landfill reports and felt an updated assessment of the landfill's impact on the surrounding air, water, and land ecosystem should be required. The last study, completed fifteen years ago, in the dark ages of solid waste management, filled only a few pages. He suggested the group demand an Environmental Impact

36

Statement (EIS). They agreed and worked closely with the Eden Prairie City Council to pressure the government to require this study.

Coalition officers also banged on doors in the neighborhoods advising new residents of the landfill's proximity and Woodlake's proposal to expand it. As they spoke with people, they solicited funds to hire a lawyer. However, with a serious recession gripping the country at the time, they met with little success. I know that after we paid our 15% interest rate mortgage each month, put curtains on our windows, and fed our children, we had no extra dollars to donate to anyone. The men felt discouraged with their failure to raise funds, but kept a combative spirit alive in the neighborhood for several years, successfully turning out citizens for meetings.

I wasn't much help in those early days. Getting through each day, caring for my children, and shoveling snow at 5:00 a.m. before the boys awoke, was more than I could handle. Ivan frequently traveled out of town, so I had to keep our home fires burning for days, sometimes weeks on end. I did somehow manage to attend a few local dump meetings, and tried to keep informed of the latest developments from involved neighbors.

I often cornered Stan Johannes at parties to fill me in on the news. In fact, I called him the "dump man," a name Ivan felt particularly unflattering and advised me to discontinue. Naturally, I thought it funny until I inherited the title of "dump queen." Somehow it didn't seem particularly hilarious anymore; in fact, it felt rather unsavory, like garbage itself.

The coalition's prime concern with Woodlake's landfill location was devaluation of our property. Though I cared about losing money on our home, this problem didn't bother us much, because we had no intentions of selling soon. I believed, like most people, the landfill posed no immediate health problems to our families, and I had full confidence in my neighbor's efforts to block any expansion with their delay tactics.

Before Woodlake Sanitary Service's expansion announcement, Eden Prairie officials learned the company had changed its mind about closing down the operation. This deception gave the city its first inkling that Woodlake's word was worth less than the trash in its dump.

Furious, Roger Pauly, City Attorney and husband of legislative candidate Sidney, headed in a fury to the landfill to check out the situation. He sensed the landfill's capacity appeared larger than permit measurements allowed, so he ordered a survey to check the numbers.

Surprise! Surprise! Roger's hunch proved correct. An engineer measured the current landfill capacity at 6,434 acre feet and not the 3,100 acre feet calculated for the original permit. Woodlake more than doubled its capacity without Minnesota Pollution Control Agency permission, a blatant violation of the permit. Incensed, Roger demanded an explanation from Woodlake.

Roger was normally a quiet man, not prone to anger easily, but he came from "no nonsense" German stock. Crossing him isn't something I'd recommend to anyone. He stood more than six feet tall and his commanding voice carried authority. Woe to anyone who became the target of his wrath. I felt pleased to stand beside Roger and not opposite him in this case.

Roger's solid Roman Catholic values were carved in granite, and Woodlake's deliberate, unauthorized expansion violated his sense of fair play. The Pauly family had a long history in the community because they were pioneer settlers in Eden Prairie. Roger raised his family in the same farm house where he grew up; in fact, he was born in the room that now served as his library.

He saw the evolution of Eden Prairie from a small farm community to a city of 44,000 people. As a child, little Roger pedaled his tricycle on the dirt road outside his home that was once an Indian Trail and is now Highway 5, a main thoroughfare serving the Twin Cities. He was the first college graduate in his family, and was now a lawyer. He'd be damned if he'd allow Woodlake Sanitary Services, Inc. to come in and wreak havoc in the city his ancestors helped build.

I didn't know Roger in 1982, but I can only imagine the confrontation that took place between him and Dick when he demanded an explanation for the landfill's illegal expansion. Dick no doubt grinned and tried to soften Roger up with some lame excuses for the "faux pas." I imagine Roger shot back his laser stare and showed no sympathy for Dick's situation.

Later, I learned from several sources the assorted reasons Woodlake gave for Flying Cloud Landfill's "surprising" capacity increase. Landfill officials suggested that perhaps their original capacity measurements were incorrect; in the old days, calculating the volume of an irregular area wasn't yet a honed science. In fact, officials said they probably wrote down the wrong volume on the original permit, and only after they remeasured the hole with advanced methods, did they discover their mistake.

Mighty big error, I thought, on first hearing this explanation. Miscalculating volume by a factor of two does little to inspire confidence in a company about to ask for another capacity increase. Maybe on the next round they'll slip up by a factor of five, and instead of beautiful prairie stretching to the horizon from our homes, we'll see only garbage!

Since their first explanation of a capacity increase didn't satisfy us, BFI's Woodlake Sanitary Services officials offered another. Bulldozer operators said they excavated the original kettle bigger to obtain sufficient soil needed to cover the garbage each day. Thus, more space became available in which to dump garbage, and Woodlake felt obliged to fill this space with additional refuse. This process probably would have continued unchallenged had Roger not dropped by to inspect the operation.

Not impressed with any of these explanations, City Attorney Roger Pauly carefully composed a six page letter to all concerned government agencies saying, "there is no question...the actions taken by BFI are in naked violation of law." In addition, Roger demanded an Environmental Impact Statement (EIS) be done for the surrounding neighborhood. He wrote, "The Eden Prairie City Council has expressed its serious concerns with respect to the potential impact upon the health and safety of the residents of the community and surrounding areas resulting from the extended operation and expansion of the landfill."

Years later, we often referred back to "Roger's letter" in our quest to shut the landfill and stop the expansion. My dog-eared, faded, xeroxed copy has been read, reread, underlined, xeroxed again, and mailed, handed and faxed to government officials around Minnesota and Washington D.C.

On seeing Roger's anger, Woodlake officials raced over to the Minnesota Pollution Control Agency to discuss the possibility of an amended permit to include their illegal expansion. Roger discovered these shenanigans and demanded that BFI complete an EIS before the MPCA issued a revised permit.

Greg Downing, Solid Waste Director of the MPCA, advised BFI

he'd take no further action on expansion until the matter of a revised permit was resolved.

Then, during the summer of 1982, two landfill monitoring wells on the southwest side of the landfill showed contamination. The State Health Department expressed concern about the problems of methane migration and groundwater flow reversal. Woodlake officials expressed shocked disbelief that their dump could be the source of pollution.

Often, landfill lawyers exhibit interesting reactions when caught in compromising situations. Utter surprise, total shock, complete amazement, and "blame the other guy" were some of the antics I witnessed. My children would attempt these same kinds of tactics on me when I caught them stealing a handful of Oreos from our cookie jar or whacking their brother's head with a toy truck. When I discovered the culprit, I'd be given a perplexed look which implied, "How dare you suspect me of such an insidious act, the darling child that I am."

I often watched grown-ups behave much like children as they tried to defend their flagrant flouting of the law with similar responses. In raising my boys, I tried to make them responsible for their actions and expected to see improvement as they grew older. Yet year after year we watched the same BFI officials give repeat performances. Perhaps they hoped one day we might believe them. It never worked. We just dug in deeper and grew increasingly determined to demand fairness for our families.

In November of 1982, Cary Cooper attended an MPCA meeting where a Woodlake official was asked what would happen if BFI continued its operation with an invalid permit. The official bluntly said BFI would shut down the facility if it failed to secure the proper permit. BFI then opted to pay for an Environmental Impact Statement hoping to pave the way for a permit. The price tag: $110,000.

January 5, 1983, marked a milestone in Woodlake's history. With a stroke of a pen, an MPCA official signed a revised permit to accommodate an unauthorized expansion of the dump, and provide an additional bonus of 10% more capacity. The reason, the official stated, was that they underestimated the original capacity. (Obviously, the MPCA believed Woodlake's first explanation for Flying Cloud's mysterious capacity increase.) Both city council and citizen's group smelled smoke from a back room deal. Neither group was present or informed of this matter until after it was done.

During the summer of 1983, the bluff washed out again. WSSI tried repairing the damage by putting a wire mesh over the slope and reseeding it. But the seed grew poorly and the slope continued to erode. Joe Mengel, a geologist and armchair philosopher who later worked with us, made these observations about the forces of nature: "There is no such thing as an impermeable rock, unchanging slope, or unaltering human institution."

The Metropolitan Council approved a scoping decision for the EIS. This study took the next several years to complete, resulting in a document 416 pages long. Monitoring wells that previously showed pollution, now, miraculously, tested clean. Incredible! I felt sure this must have been a sign of divine intervention. Perhaps the pope should be notified of this other world phenomenon. Maybe the good Lord chose to make clean this toxic land and purify the contaminated water, thus restoring the earth to its former splendor as a Garden of Eden.

Dr. Roy stated if there was any leakage from the landfill into groundwater, it would be significantly diluted before reaching the groundwater. We filed his comment, along with his recent lab test results, in our mental garbage cans.

The EIS, commissioned and paid for by BFI's Woodlake Sanitary Service, Inc., found no major adverse environmental impact from the landfill expansion. I always found it curious a company being monitored should study its own pollution. We knew Woodlake wouldn't admit to problems with its operation and sabotage its hope for an expansion.

During the state legislative session that year, Woodlake slapped an amendment onto a lengthy waste management bill to secure an exemption from a Certificate of Need required for other state landfills. The bill passed.

On December 27, 1984, the Metropolitan Council declared the EIS final and adequate. City officials argued the document did not address potential environmental harm, public liability, the need for the project, and it's inconsistency with the Metropolitan Council's call for landfill abatement. Unfortunately, the government didn't share the city's point of view and voted to approve the study. The matter moved forward to the MPCA.

In January, 1985, MPCA officials began drafting a landfill expansion permit in what government officials described as a "give and take" exercise. I had a hard time wondering what Woodlake gave up in these sessions. Up until now, the only legacies WSSI gave the community were broken promises, misleading data, polluted drinking water, frequent fires, garbage in the river valley, and so on.

During the 1985 legislative session, our local and state politicians tried in vain to repeal Flying Cloud's exemption from the Certificate of Need. Christmas was over and Woodlake was not in the mood for gift giving. Eden Prairie officials failed to win their case. "They got nine silk-suited lobbyists to defeat the amendment," stated an irate state Senator Don Storm in one of his later speeches.

In May and June of 1985, the Eden Prairie Planning Commission and the Eden Prairie City Park's Commission had recommended that the City Council deny Flying Cloud Landfill's request. They said an expansion would be incompatible with adjacent neighborhoods and the airport, environmentally unsound, and appeared to exceed the county's need for landfill space.

When Woodlake did its normal well testing, contamination was rediscovered. Wells were tested again a few weeks later and the data showed unsafe levels of toxic chemicals in the water. So much for miracles! I threw out my letter to the pope. With the permit process underway, Woodlake officials must have figured, "Why rock the boat?" They neglected to report their findings.

Meanwhile, Woodlake's expansion permit went before the Metropolitan Council and received a 9-7 vote of approval. July 2, 1985, Dick appeared before the city council to request the expansion. There was no mention of well contamination here either.

On July 8, two months after pollution was detected in the water, Woodlake submitted the damning data to the Minnesota Pollution Control Agency (MPCA). Government officials suspended consideration of the expansion until a remedial investigation was undertaken to study the extent of pollution and to propose remedial action. Eden Prairie got the word, and City Attorney Roger Pauly hit the ceiling of City Hall. The news made the Minneapolis *Star Tribune*.

August 25, 1985, Flying Cloud Sanitary Landfill earned a spot on the State Superfund List. It deserved placement on the Federal Superfund List also, but since BFI said it would clean up the water contamination, the MPCA let them off the hook once again.

Unfortunately, now, most members of the Eden Prairie Citizen's Coalition (EPCC) had burned out, moved away, or been recruited to the planning commission, resulting in a conflict of interest. Seeing this fall in citizen fervor to fight the expansion, the city considered a compensation package for the neighbors Woodlake offered. The remaining members of EPCC supported the idea, hoping to receive a large enough monetary settlement to offset diminished property values if the expansion were approved.

Cary Cooper explained to me later that he and his colleagues "just couldn't keep pace with a process that blatantly favored our opponents." EPCC leaders couldn't match the time, money, and energy of Woodlake's full-time paid lawyers and lobbyists, so they disbanded.

With the EPCC defunct and Eden Prairie officials willing to accept a bigger garbage dump for money, BFI's express train now roared down the tracks to its final destination: an expansion worth millions of dollars. But could we afford the price of a one-way ticket? The cost: possibly the health and safety of our loved ones.

THE POLLUTERS

3 | RING AROUND THE DUMP

ebruary 19, 1986, I ran outside in my bathrobe in the frigid early morning air to retrieve the newspaper. Back inside my warm home, I settled down to the day's top stories and a breakfast of tea, toast, and cheese. As I paged through the paper, headlines on page six of the Metro section grabbed my attention. "Contaminants Found in Groundwater Near Landfill." My heart pounded. I continued to read, ignoring my breakfast. Environmental reporter Dean Rebuffoni wrote, "Chemical contaminants from the Flying Cloud Landfill in Eden Prairie have seeped into groundwater outside the disposal site's boundaries." That was enough. I shut the paper.

My thoughts now flashed back to nightmares and fears I had growing up as a teenager in Cresskill, New Jersey. When I was fourteen, my five year old sister, Karen, died of leukemia. The local pediatrician, Dr. Ferber, believed buried hazardous waste might have caused her illness.

Cresskill served as a World War I training camp where soldiers used mustard gases and other toxic chemicals preparing for combat overseas.

45

Years later, remnants of these materials resurfaced when someone's spade hit them while putting in a garden or swimming pool. Since the area had a high incidence of childhood leukemia, townspeople thought there might be a link to the buried waste.

Watching my little sister die and seeing the torment on my parents' faces as they buried the youngest of their five children, disturbed my tranquility for years. I can still hear my sister's screams piercing the normally silent hospital corridors on the day I visited her so long ago. The doctors were administering a painful test.

Losing my sister stole my youth. It brutally forced me into the world of adulthood where people don't always live happily ever after and rainbows don't follow every storm.

Our family was never the same after Karen's death; we worried that we would also become victims of leukemia, or that our own families would later succumb. I worried now that history might indeed repeat itself. A feeling of "deja vu" soaked my soul.

I doubted whether I could survive the agony of watching my own flesh and blood precede me in death. Children dying before their parents defies nature. A young boy or girl cut short in youth, never to reach full bloom, violates one's sense of justice. Having seen the pain of this tragedy firsthand, I had no desire to relive it.

Fear compelled me to get more involved in the landfill issue. Intuition led me to believe an expansion of a polluting garbage dump could impact more than just our property values. Human life might be jeopardized as well.

In my emotional state, I tried evaluating the situation logically, drawing from my biology studies completed eons ago. According to my children, that was "back in the old days when hippies and rock and roll were groovy."

I remembered learning that the young and old are the weakest members of society and, therefore, most susceptible to disease. The young absorb toxins rapidly because they metabolize food and other substances quickly. Children have fewer detoxifying enzymes, breathe more air for their body weight, absorb toxic chemicals more readily, and have cells that divide more rapidly, making them more prone to cancer.

On the other hand, older folks repair damage to their bodies at a reduced rate. This makes them equally susceptible to serious illness resulting from contamination.

My four-year old son Paul fell into the younger classification. He was now one year younger than my sister when she died.

In late August, 1985, Louis Affais called a meeting at his home for anyone concerned about water contamination. John Kennedy, our neighbor across the street, and I walked the two blocks to Louis' home to find out about the situation.

The new faces around the Affais' kitchen table represented fresh blood for the cause. They also meant that we faced a steep learning curve in educating ourselves on the issue before we could take effective action. What we lacked in knowledge, however, we compensated for with enthusiasm and optimism. Back in those days, that was the grand total of our assets. The liability column on the other hand was lengthy. It included no understanding of hydrology, geology, biology, chemistry, or landfill history, expansion, and impact on surrounding areas, no political savvy, no knowledge of environmental laws and agencies involved in the permit or licensing process, no lawyer, no money, no organization, no friends in the right places.

Of course BFI's Woodlake Sanitary Service commanded complete mastery of the situation; they had millions of dollars to hire lawyers, scientific experts, and lobbyists; and they knew the state political scene for years, and had cultivated cronies in high places. In the end, however, our iron will to protect our families made up for our initial deficits.

As I reminisce on our innocent beginnings, I figure our chances of defeating a landfill expansion were about .0001%. More than a few people thought we were chasing windmills like Don Quixote, in pursuit of an impossible dream. We were young, naive, and believed with all our souls that reason and justice should prevail in a democratic society.

Thank God we lacked power to see the future, or we might have run away, fled the country, joined a cloistered order of priests or nuns, or done all three had we known about the troubled path ahead of us. This day at Louis' house, we took our first steps on "the road less traveled," beginning a journey we were poorly equipped to make.

Louis served us pure orange juice, an environmentally correct thirst quencher, as we listened attentively to Jodi Thomas of Clean Water Action tell us we had little hope of holding back the momentum of this garbage giant. There was nothing she could do, nothing for us, she said.

And probably none of the mainstream environmental groups would touch this case either. It was too political.

The only help she offered was the name of Leslie Davis of Earthprotector, an environmentalist who represented a more radical group. Jodi advised us to call him and see if he'd support our quest. Since we had no other options, we agreed to do this.

Earthprotector leads the charge

A grinning, black-on-black attired man stepped through Mary Kennedy's doorway one September evening and introduced himself as Leslie Davis. His five-foot, six-inch frame sported a black Earthprotector T-shirt worn over black rumpled trousers. A black beret sat atop his black hair, and black and white high top sneakers completed the color coordinated ensemble. Framing his pale face was a salt and pepper beard, giving him a Middle Eastern air.

We wondered whether Leslie's attire reflected Paris' latest black sportswear collection or the trappings of a guru beckoning us back to a simpler lifestyle. We later learned that he dressed in the prescribed color of his Jewish faith and concealed a yarmulke beneath his french beret. He also served as President of his own company, Earthprotector, which sold apparel and used part of the profits for environmental work.

After the group recovered from his appearance, we said "Hello," and he responded with "How wa ya." I recognized the New York accent because I grew up nearby in New Jersey. Later, I learned Leslie had lived as a child in the same neighborhood where my father grew up.

Seating himself on the church pew in Mary's family room, Leslie happily preached to us about the political maze through which the expansion permit must travel before it became a fait accompli. We learned that the Metropolitan Council would study the case again, and that the Environmental Quality Board and Pollution Control Agency must also give their stamp of approval. A Supplemental Environmental Impact Statement, a Remedial Investigation, and Feasibility Study still needed to be prepared. Unbelievably, this was just the beginning! We begged Leslie to stop; our brains suffered from information gridlock.

If a landfill literacy test were administered at this moment, we would have scored in the lowest ten percentile. Industry buzzwords defied our comprehension and the process resembled an obstacle course. Then there was the landfill data, another mass of scientific information that needed to be ingested and digested before we could testify intelligently before various governmental authorities. I believe 90% of Leslie's evening sermon sidetracked my brain and fell out my ears, landing on Mary's orange carpet.

Late that night, after I had wearily bade goodbye to my fellow dump-busters, I slowly walked across the street to my house with head down and shoulders hanging. The task before us overwhelmed me. It wasn't long before I complained to Leslie on the phone. I felt helpless trying to climb the gigantic, steep mountain that lay ahead of us. He responded gently saying that "some people don't even have the courage to walk to the foot of a mountain and look up." That remark raised my spirits somewhat, but the mountain still remained to be climbed. Leslie's words gave me the strength to take the first steps.

Double, double toil and trouble

As I began plodding through landfill reports, my brain felt like a dead battery badly in need of a jump start. I wondered if all those years I stayed home caring for my children had made me incapable of memorizing more than "Wee Willie Winkie runs through the town," or "One two, buckle my shoe." Didn't I earn a college degree in biology which required that I memorize the Krebs Cycle, a formula that occupied an entire loose leaf notebook page? I could even conjugate the verb "to be" in all it's tenses for my French III class. *Que c'est il passe?* What happened?

Despite misgivings about my mental acuity, I resigned myself to the task of learning what took place at the Metropolitan Council and Minnesota Pollution Control Agency, and who the key players were. These were the most important governmental hoops through which the permit must pass on its travels down the bureaucratic pipeline.

We learned that the Metropolitan Council plans airports, roads, landfills, and so on in the Twin Cities, and the Pollution Control Agency writes and enforces regulations that comply with Federal pollution laws. We requested reports from both agencies, then proceeded to read these piles of papers, attempting to comprehend the history, biology, geology, hydrology, and chemistry of the Flying Cloud Landfill. Being an equal opportunity organization, we exchanged volumes of studies between

49

households, giving all concerned residents the chance to discover for themselves the latest round of volatile organic compounds contaminating our drinking water, and the spread of the leachate plume into our neighborhood.

After several months of study, we became landfill literates. Leachate, we discovered, is the contaminated water discharged from a landfill after rain washes through the garbage. The area of groundwater contamination is known as the plume, and according to the latest hydrogeological maps, it was spreading in our direction and flowing south into the wildlife refuge. At present, it lay within 150 feet of occupied homes.

Though a treatment plant processes water before it runs through our faucets, it doesn't routinely test for many contaminants. Unfortunately, several nearby residents and the local drive-in theater were pumping up contaminated well water for immediate consumption. Later, BFI supplied them with bottled water.

BFI's own laboratory analysis showed that the aquifers contained an alphabet soup of chemical compounds. Acetone, benzene, 2-chlorethylvinyl ether, methylene chloride, toluene, trichloro-fluoromethane, vinyl chloride, and xylene are just a few of the dozens of assorted chemicals present. Many of these compounds, such as methylene chloride and ethyl ether, are solvents. Others are used in manufacture of pesticides (1,2 dichlorothane) and PVC (vinyl chloride). Some, like benzene, are carcinogens which cause cancer.

Even very low levels of some pollutants can be detrimental to one's health if they're consumed daily in tap water. Vinyl chloride is one such compound that has acute and toxic effects on humans in concentrations of just 1 part per billion. This is equivalent to one bad apple in 2 million barrels. One monitoring well at the landfill showed 74 parts per billion of vinyl chloride. Now that's a lot of rotten apples to sicken the population.

Another compound, dichlorodifluoromethane is a refrigerant, present in quantities up to 2000 parts per billion in water samples. With numbers like these, untreated groundwater easily failed Federal drinking water standards.

We reasoned that if chemicals could migrate through sandy soils into groundwater, then they could just as easily drift through the air. But that wasn't such a brilliant deduction on our part. Noxious odors emanating from the dump on warm, still, summer evenings constantly reminded us

50

of our unwelcome neighbor over the fence. Of the many organic compounds contained in the landfill, some volatilize when exposed to air, and many volatile organic compounds (VOC's) are also carcinogens.

"Ten per cent of the time wind doesn't blow in the direction of Homeward Hills," said one BFI official in a memorable statement.

Being fairly astute at arithmetic, we calculated that 90% of the time the wind hit us with a mixture of VOC's. When the air hung heavy over the neighborhood, those people living closest to the landfill shut their windows some evenings and stayed indoors to avoid breathing the chemicals.

State law requires that garbage be covered by soil to prevent litter and noxious odors from escaping, and to prohibit rats and birds from feeding on garbage. Judging from neighborhood air quality and the number of childhood bronchitis cases, we doubted whether landfill officials consistently obeyed this regulation.

Many other questions worried us. For example, how did high levels of toxic chemicals like refrigerants and solvents find their way into a municipal landfill designated for household waste? Could the resultant, nauseating foul air make us sick? Why didn't government officials appear concerned with this situation? Why didn't they order an investigation? My initial frustration with so many unknowns prompted me to label the landfill, "The Great Mystery Pit."

The bubbling toxic stew in the landfill kettle reminded me of the witch's boiling caldron in William Shakespeare's play, *Macbeth*. It contained anything and everything, and brewed together for years to become a potent mixture. I caught myself repeating the witches chant as they stirred their bubbling cauldron in the forest:

> *"Double, double toil and trouble;*
> *fire burn and caldron bubble.*
> *... For a charm of powerful trouble,*
> *like a hell-broth boil and bubble."*
> *"Double, double toil and trouble,*
> *fire burn and cauldron bubbling."*

Next we studied methane gas, the natural by-product of organic decomposition produced in massive quantities at landfills. It's odorless, colorless, lighter than air, and seeks the path of least resistance to escape into the atmosphere. Mixed in concentrations between 5 and 15 percent in the air, this gas is flammable at atmospheric pressure and ordinary temperatures.

Everyone knew horror stories of methane gas migrating into basements and exploding homes. Since subsoils of the Flying Cloud Landfill site were porous, they provided an excellent medium through which methane could travel laterally. One neighbor explained to us that methane can also seep through spaces along a gravel-packed pipeline and enter houses this way. He was familiar with the problem, because little flags extending in a line between his home and the neighbors marked the location of the natural gas trunkline serving his subdivision. This pipeline also ran adjacent and parallel to the landfill.

"We have several options for blowing up the neighborhood," I announced with feigned glee at one of our zillion neighborhood meetings. No one found this joke particularly amusing but me, because recently the Minneapolis *Star Tribune* ran a story about a gas pipeline explosion in a nearby town. Disaster seemed inevitable as we gradually unveiled the mysteries shrouding the great garbage pit.

The bureaucratic jungle

After we researched as much as we could on our own, we contacted governmental agencies for additional information. Our first call went to John, technical expert at the Metropolitan Council, who knew the "ins and outs" of testing procedures at Flying Cloud Landfill and could interpret data for us. In preparing for the first showdown before his agency, we took turns picking his brain.

On one occasion I confided to John my utter dismay that Woodlake be allowed to contract with the local Pace laboratory to do its own testing. He assured me this is standard procedure and explained that MPCA splits samples with Woodlake, on occasion, to check results. However, the company normally reports its own findings. Even after Woodlake withheld sampling results, the government continued to allow it to monitor its own business. State Senator Don Storm likened the situation to "the fox watching the chicken coop."

John played a pivotal role in the landfill permit process. He testified at Metropolitan Council, MPCA, and Eden Prairie Council hearings, as well as for the state legislature. And he took a beating in this case. I believe he was tormented between speaking the truth and trying to keep his government job. We suspected he was pressured to provide technical data to justify the landfill's expansion. However, as a father of two children, he must have understood our sentiments. No one envied the dilemma in which he grovelled for years.

Scientifically, there was no middle ground. The Mayor later characterized the situation well when he eloquently testified before the Metropolitan Council saying, "a little rape is still rape." The unlined, leaking landfill will always pollute drinking water, and both a little rape and a little pollution are wrong.

Through my many hours on the phone with John, I learned his life's story and felt we established a good rapport. But he betrayed this trust several times, trying to compromise the truth with what I suspected he felt obliged to do and say. Despite these slip-ups, John's young charm reminded me of John Boy, in the television show, *The Waltons*, so we anointed him with this nickname.

One day, I accepted John's invitation to peruse the gargantuan, *Environmental Impact Statement* in the Council building downtown. As I sifted through volumes of information, one fact in particular jumped out: damage to our nearby wildlife refuge would be irreversible. I repeated this sentence over and over as evidence that the fragile ecosystem endowed to Eden Prairie could be lost if it became too badly contaminated.

We contacted Ken Podpeskar, the engineer assigned to our case at the Pollution Control Agency. He also patiently discussed any questions we had. When I frequently spoke with him over the phone, however, my three boys always seemed to be engaged in hand-to-hand combat. So our phone conversations were constantly interrupted with my threats to send them all to their rooms until they were old enough to enter college. Ken would just laugh. He always recognized my voice on the phone because the background noise consisted of loud fighting.

In time, my perpetual presence on the telephone with neighbors, politicians, government officials, and lawyers, began irritating the children. The situation got worse. Eventually, our telephone rang every day nonstop from seven in the morning until eleven at night. Before long, my children cruised the neighborhood spreading rumors like, "Our

Mom is going to have an operation; she's getting the phone cut out of her ear."

The guilt I felt for taking time from their care hung heavy. We tried ignoring the ringing phone during mealtime and bedtime, but I hated missing important calls. Finally, Ivan came up with a solution. He marched home one night with an answering machine tucked under his arm and plugged it into the phone. From that time on, he refused to take calls during our family time.

Cement boots In Purgatory Creek?

Back in those bad old days, I often hopped on my bike and rode to Louis's house to talk out my latest worries about fighting BFI. We sat together on his front porch and soul searched late into the evening discussing prospects of taking on the world's second largest garbage company. Had we gone mad, like people you read about in the newspapers who do bizarre things? We wondered.

Furthermore, we read in the newspapers, like everyone else, that garbage and organized crime often go together like peanut butter and jelly. Was this the case with BFI? "We might end our days wearing cement boots in Purgatory Creek," Louis remarked. Hysterical at the clever joke, we laughed long and loud. In fact, Purgatory Creek is a real river that meanders near our homes and behind Pax Christi Catholic Church. It provided us with an endless source of amusement. We jokingly plotted to throw a BFI lawyer over the bank for his sins against us.

I'm not sure at what point we made the conscious decision to go full throttle with our mission to shut down the landfill. Gradually, as it became clear to us we weren't living next to a "sanitary" landfill, but an environmental time bomb exploding out of control, we became increasingly involved.

We had initially hoped that presenting our findings to the government would convince officials not to expand an already leaking garbage dump in a residential neighborhood, adjacent to a wildlife refuge, over three drinking water aquifers, on the edge of an unstable bluff, and next to an airport. Approval of expansion would demonstrate solid waste mismanagement. Surely the government would see the light when we turned it on. Surely officials would put health and safety interests of citizens before corporate profits. How terribly naive we were back in the bad old days!

Homemaker rebellion

As we gained confidence handling facts in the case, we geared up to spar with experts. At the recommendation of Leslie Davis, we requested an information meeting with the MPCA to bombard them with questions about soil and groundwater contamination. We hoped to impress agency officials with our newly acquired landfill lingo, and expected to punch holes in their technical arguments.

Mary Kennedy and I wrote our first flyers to drum up support for our cause. As neighborhood stay-at-home Moms, Mary and I would eventually spend an inordinate amount of time together designing more flyers, attending more meetings, making more phone calls, organizing more neighbors, setting up an association, raising money, testifying before groups and so on. A vivacious 30-year old, Mary had a knack for public relations and organizing. Her help was invaluable in those early days. Patti Vining, a young mother on Jack Pine Trail, agreed to typeset flyers at her office. The results impressed us, so we congratulated ourselves on a job well done.

Over the years, many more mothers joined our forces and we became the front lines in this battle to safeguard the neighborhood for our children. Maternal instincts to protect our young drove us out of playgrounds and backyards, into Minneapolis and St. Paul to testify before the Metropolitan Council, MPCA, courts, and the state legislature against a dump expansion we believed could harm our children. Our conversations switched from toilet training techniques and the current strain of flu hitting the schools to VOC levels in groundwater and movement of the leachate plume closer to our homes.

We enlisted recruits to our cause at PTA meetings, hockey rinks, soccer fields, the school carnival, and wherever else people gathered in town. A local garbage man once complained to a friend of mine, "Those women living near the landfill have nothing better to do than sit around the coffee table and plot to shut the dump." He was right. We put our lives on hold so we could attend meetings and spend off-hours glued to the telephone, rallying support and developing fresh strategy.

We became like possessed creatures, defending our children in much the same manner I'd seen a mother bird fend off Killy, the Varlamoff family cat. Once Killy quietly crawled over to a happy nest of robins, preparing to feast on tender babies for a gourmet luncheon. On seeing the threat to her defenseless offspring, Mama bird, outsized by this feline, attacked with all her strength. Then she chirped an urgent SOS to all her feathered friends. Quickly, other females joined the assault. Not wanting to deal with these hysterical mother birds pecking at her, Killy spun on her paws and headed in the opposite direction. Store-bought cat food would just have to do for that day.

Since our landfill fight occupied so much time, we decided to bring our children with us to meetings as often as possible. This served the dual purpose of keeping them close by and reminding all concerned parties that the neighborhood swarmed with hundreds of running, bicycling, and skateboarding children, ranging in age from tots to teens. Thus, dump crusading became a family affair.

At meetings, babies and toddlers in strollers amused themselves with car keys while four- and five-year olds crayoned in coloring books next to their mothers and fathers. And there were always several eight- to ten-year old boys scuffling in the back of the room, often led by my son Pierre. The presence of the children kept our focus. It was for them that we walked for years through raging fires of difficulties to battle infinitely greater forces than ours. It was for them we consecrated this fight.

Our sons and daughters would grow up to become landfill experts, too, receiving their education at the hundreds of meetings we attended over the eight-and-one-half year crusade. In time, five- and six-year olds discussed ground water contamination and understood the acronyms, VOCs, BFI, MPCA, and EPA. Their lives would never be the same after this experience. We hoped this environmental indoctrination would last a lifetime and be passed on to their children's children.

A duel with experts

The neighborhood emptied once again to attend the Minnesota Pollution Control Agency (MPCA) information meeting at Hennepin Technical Center on October 23, 1985. Mary and I stood in the doorway to greet several hundred people who came. As usual, BFI officials led the meeting, droning on about technical merits of their "state-of-the-art sanitary" landfill expansion.

56

BFI's lawyer, Dick, once again gave his usual, grinning, hair-stroking performance. His slick style prompted us to call him "Tricky Dicky" since we felt sure he was slipping something by us with a smile on his face. However, he abandoned his golf course plan, sparing us that aggravation for the evening and making me very glad I didn't buy clubs.

Dick's sidekick, Dr. Roy, flew in from Chicago again. In his introduction of himself, he pointed out that his last name rhymed with baseball. So "Baseball" became his undisputed nickname in Eden Prairie dump-buster circles, and as usual, he bored us with his monotonous explanations. He also mentioned he believed landfill contamination stemmed solely from chemicals contained in ordinary household waste. "Moan!, groan!," went the crowd. Another entry for our mental trash bin.

A sense of humor went a long way towards relieving the stress of crusading against a Fortune 500 company. During long tedious hours of public presentations we often poked fun at the manner and dress of our opponents. I agree with Lord Nelson who said, "I could not tread these perilous paths in safety if I did not keep a saving sense of humor."

Since "Tricky Dicky" changed his hair color and style frequently during the battle, we couldn't resist teasing him about his latest "hairdo." The color ranged from sandy to blonde, to salt and pepper, to white, then back again to sandy. When Dick ran his fingers through his hair, it often stood on end giving him an electrocuted look after testifying. (I am sure he must have felt that way after dealing with us.) Eventually, he abandoned his hair stroking performance for a "hands in the pocket" routine. Maybe Mary Kennedy's mimicking of him convinced Dick to switch techniques.

I believe this case so frustrated him, he could have torn out his hair completely. Before it all ended, Dick's hair turned snow white. He changed law firms and finally withdrew from the case.

After BFI's lawyers and experts usurped prime meeting time for their presentations, we were left with the last remaining minutes for ours, before the building was shut down. Also, half the audience had fallen asleep. This was standard meeting procedure until we realized we were slowly being squeezed out.

Unfortunately, our spokesperson, Leslie, gave a poorly prepared slide show to start our presentation. Many slides appeared upside down on the screen, embarrassing us. Leslie, however, was quick on his feet and commented that his show represented the state of affairs in the government. The crowd howled their approval.

A fellow homeowner in our group stood up next and asked a plethora of questions. He did his homework well and had MPCA experts admitting they didn't have all the answers. Then his young son interrupted the discussion to inform him that it was his bedtime and mommy was taking him home to sleep. "Goodnight, Daddy" he said. Again, an important human dimension kept us aware that this was not just a case of abstract facts. Young lives were at stake, too.

Lacking the confidence to speak out publicly at this point, I added nothing to other citizen's remarks. I grew up a very shy child, coercing my more outgoing sister and brother to do my bidding for me. Speech class in high school and college made me tremble. It frightened me to be the focus of attention in a large group of people, so I convinced others to speak out instead, and wildly applauded their remarks.

Our First Protest

We continued monitoring dump developments throughout the fall of 1985. The Metropolitan Council ordered a Supplemental EIS to assess the impact of a leaking Flying Cloud Landfill on its immediate environment. Woodlake, confident this study would reveal no problems, requested permission to excavate a hole for the expansion. The MPCA denied approval.

Soon after, landfill neighbors noticed 10 earth movers pushing dirt around the facility. They relayed this information to the city manager, and he reminded BFI they were forbidden to dig. Sidney Pauly also notified the MPCA that BFI was excavating for the expansion without approval. An official shot back that "MPCA has bigger fish to fry than some landfill operator who is moving dirt around." (It was hard to believe there were many bigger fish in Minnesota than this 3 billion dollar one!).

By late January, 1986, Woodlake was brazenly bulldozing an area for expansion, and Eden Prairie officials obtained a temporary restraining order to stop them. Newspapers quoted BFI's lawyer, Dick, as saying he was frustrated. The city wouldn't give BFI permission to dig for an expansion, so the company went ahead with its plans anyway.

Looking for a sympathetic ear, Sidney Pauly went to discuss the

58

incident with Metropolitan Council Chair, Sandra Gardebring. Furious, Sandra declared that the Metropolitan Council would join the city's lawsuit against BFI's Woodlake Sanitary Services, Inc. The city won the case, and excavation at the landfill was prohibited until proper permits were issued.

On May 1, 1986, Woodlake applied to the Environmental Quality Board for a variance to expand its landfill and awaited approval of a permit. They had opted for above-board tactics this time. The stated reason for their request was economic hardship. Feeling no pity for their plight and concerned they might succeed with this strategy, we planned our first demonstration to stop them.

Leslie envisioned a protest with thousands of citizens walking in solidarity around the landfill to draw attention to its proximity to the neighborhood and wildlife refuge. He called it "Ring Around The Dump." It was a wonderful idea, but we knew nothing about preparing for and carrying out demonstrations. Though most of us were 60's children, we weren't the bra burners and college campus protesters of that era. Since Leslie knew the way, we followed him.

Timing was important. Leslie designed flyers showing people holding hands around a garbage can and chose Sunday, June 1, 1986 for the event. This was the day before the Metropolitan Council was to decide the fate of the variance.

Next Leslie sent out press releases and called a press conference at the dump to state his intentions. I accompanied him and received my first exposure to being interviewed by reporters. I was nervous, of course, and let Leslie do most of the talking.

As June 1 neared, eight to ten people met every Sunday evening at my house to finalize preparations for the demonstration. It was easier having meetings in my living room because Ivan traveled often and hiring sitters every week cost a fortune. As the evenings wore on, my son Paul, then four years old, would crawl into my lap and fall asleep. Neither the discussions nor the laughter disrupted his slumber. Paul grew up in the shadow of this dump quest. He was one year old the day we received BFI's letter, and nine-and-one-half the day it ended.

Prior to our protest walk, we invited other members of the community to join in the preparations. Twenty people attended a meeting at Pax Christi Church and offered many good suggestions which we adopted. One person proposed a helium balloon launch, and someone else recommended that participants sign a petition we could later present to the government. Another neighbor proposed renting a megaphone for crowd control and instructing walkers to hold hands in a show of solidarity.

Doug Larsen, columnist for the *Eden Prairie News* volunteered to lead the procession carrying the American Flag. He believed our demonstration showed the American spirit for independent thought. In his weekly column, he had lambasted Browning Ferris Industries (BFI) for calling their leaking, toxic landfill "sanitary" and labeled it a "dump" instead.

After we felt satisfied with our plans, Leslie divided 7000 flyers among us and we went off to canvass the city. We walked the streets, malls, soccer fields, schools, and went door to door in dump neighborhoods depositing handbills. Churches put them in their Sunday bulletins. The Eden Prairie City Council even passed a resolution supporting the demonstration. With preparations completed for June 1, 1986, we prayed for sunshine.

Our wish was granted. Sunday dawned a clear, sunny day. A slight breeze blew across the bluff and the temperature rose to 70 degrees before stopping. We felt destined for success. The media called Leslie early in the day requesting the name of a spokesperson for the citizen's group. No one volunteered, so by default, I accepted the position.

One hour before our start-up time, rally organizers filled helium balloons and set up a booth to hold miniature garbage cans for contributions and pads of paper for petitions. Leslie arrived with a contingent of young boys from his temple, all wearing yarmulkes. They quickly unfolded some tables they brought from the temple and disappeared just as fast as they had come.

TV reporters arrived early and asked for an interview with me at my home. I brought my children and Mary Kennedy's daughter back with me to Jack Pine Trail, hoping cameras would catch the children playing in the background, while I talked about the polluting dump in our neighborhood.

A reporter handed me a microphone to slide down my "Stop Acid Rain" T-shirt, and I bungled the operation. Thoroughly embarrassed, I fixed the microphone according to precise instructions of the reporter and watched the camera zoom in on my face. I could feel myself turning beet red in the sunshine as I tried to steady my voice.

"A multibillion dollar garbage company wants to expand their toxic landfill in our neighborhood, and we feel victimized by their plans," I began.

With no time to mull over my remarks, I raced back to march around the landfill with the others. Several hundred people were already ahead of me, just out of sight. As I ran through thigh-high prairie grass, I soon caught a glimpse of them. Moving like a giant python over contours of land, they undulated up and down. As I caught up to the tail of this oversized reptile, I heard the laughing voices of friends. Then the American Flag, held high in the breeze, fluttered into view, directing the body of protesters, aged one to seventy-five, around the landfill.

I saw mothers carrying infants in their arms, and dads supporting toddlers on their shoulders. There was an old woman hobbling along on swollen ankles and a handicapped man pushing himself along the uneven path in his wheelchair. The mayor and a City Councilman were working the crowds, and a local minister even joined members of his congregation to support the cause.

I excitedly traveled up and down the line talking to everyone. My friend Carol Culp was there with her four children, and I took a turn holding her baby while I caught up on family news. Later, I saw the mayor and strongly advised him against considering any monetary deals with Browning Ferris Industries (BFI) in exchange for the dump expansion. "People won't tolerate it," I cautioned him.

Soon the procession of people came to a washed-out area of the river bluff. The path was difficult to cross, so a smiling Leslie stood there gallantly in the sunshine, extending his hand to help people pass safely. His gesture was well appreciated by mothers with young children. This inconvenience reminded us that forces of nature were working steadily to erode the bluff.

Moving onward, we came to a place overlooking the refuge. Here the group halted in unison as if hypnotized en masse. We stood mesmerized by the glorious view and paused to dwell on its opulence.

Shades of aquamarine and emerald green stretched before us to the horizon punctuated only by the muddy Minnesota River as it meandered between clustered trees. Grass Lake and Blue Lake reflected the sunlight like diamonds. Tall grasses bordering the refuge swayed in the breeze. Occasionally, the tranquil scene was broken a bird, perhaps a trumpeter swan or a blue heron, that flew heavenward.

Nature was generous to Eden Prairie. It bequeathed a treasure chest of jewels to the land, gifts to enrich the human soul.

Could this fragile ecosystem survive continued pollution from a thirty-year dump expansion? Lest we find out what this great contamination experiment might yield, we continued our walk with renewed vigor. We were determined to protect our rich inheritance.

As we circled the landfill, we noticed grim faced guards positioned every 100 feet or so on the other side of the fence, carefully watching our progress. If BFI was expecting trouble makers, they would be disappointed. We came only to appreciate the bounty of the land and make a statement.

Nearing the end of our journey, the megaphone announcer instructed all of us to hold hands while T.V. cameras recorded our show of solidarity. The day felt like a step back in time. We laughed as we reminisced about the 60's, mini skirts, love beads, and peace marches. When the protest ended, we dispersed to our homes, pleased with our demonstration.

The core planning group returned to my house and turned on the television, hoping to catch a glimpse of protest. It must have been a slow news day, because our "Ring Around The Dump" was the second item on the evening news. We cheered this success. Even my inhibited attempt to present the citizen's position sounded decent. Just then the telephone rang.

"Hi, how'd it go today?" came the familiar voice of my husband. He had been taken away on business, and now I enthusiastically filled him in on every detail.

What a great day! People of all ages, sizes, shapes, colors, economic situations, and religious persuasions came together to protest the ravaging of their land. It was a picture perfect setting with an upbeat mood that produced an experience I will always remember with joy. Sleep came easily that night.

Several days later, Dean Rebuffoni wrote in the Minneapolis *Star Tribune* that the Metropolitan Council voted "nay" to Woodlake's variance request. He went on to say that they recommended the landfill not accept commercial garbage after October, when capacity would be exhausted under the present permit. Only private vehicles would be allowed to use the facility now. Lastly, the permit process was halted until results from groundwater studies were completed. It was a stunning victory!

THE POLLUTERS

4 | MEETINGS, MEETINGS, AND MORE MEETINGS

O ur little neighborhood group was soon headed toward burnout, like the Eden Prairie City Council before us. We needed fresh troops. Also, we desperately required funds to pay for flyers, mailings, perhaps a lawyer, scientific experts, and newspaper advertisements.

For once, luck nodded our way. Economic recession lifted, and a building boom hit the area. Houses rose like mushrooms overnight in developments along Homeward Hills Road providing us enormous potential for increasing our numbers and bank balance. Homes going up were not low income models either, but custom built $500,000 splendors.

Dream Street, with its picture perfect homes containing every creature comfort a yuppie heart could lust after, grew up one mile from the dump. We felt sure that when these financially endowed people moved in and discovered they shared the neighborhood with a landfill, they'd happily commit themselves and their resources to the cause. A polluting garbage dump down the block does nothing to enhance one's address.

65

After a brainstorming session, we determined that we needed a legitimate organization from which to solicit members and raise funds. Six of us, Leslie, Louis, Tim, Mary, John, and I, unanimously agreed that a homeowners association would be the appropriate vehicle. My family's financial consultant, Terri O'Brien, had formed such an association in his Burnsville neighborhood to successfully tackle environmental problems, so I examined his organization's framework. I studied it over the summer of 1986 and wondered if a similar framework could help in accomplishing our goals. At the end of August I called together our dwindling group to present the results of my work. "The Burnsville association can work for us," I announced.

Together we changed and revised the by-laws to our satisfaction and gave our organization a name, The Homeward Hills Homeowners Association. We selected this name for the main road that joined developments east of the landfill. This area represented the highest population density near the dump. Since the name quickly became an abominable tongue twister, we subsequently shortened it to HHHA.

On Louis' suggestion, we agreed to form a five member executive board to run our association. This avoided concentration of power in one individual and gave each development representation in the true spirit of democracy. We also thought it a good idea to give members an opportunity to work on other issues they felt could improve the quality of life in their neighborhoods. We sensed that shoving our own narrow agenda down their throats might be poorly received.

To defray printing and postage costs, we decided to request a $10 membership fee, then planned a wine and cheese party to kick off the first meeting. Even if people weren't interested in our agenda, we figured they might consider a night out to party. Once we had them, we planned to wine and dine them until they agreed to join our fun-loving, dynamic organization. (Any person of sound mind should have known better than to hook up with our group of young renegades.)

By now, Mary and I had canvassed the various neighborhoods near the Flying Cloud Landfill neighborhood by designing and distributing flyers, and we volunteered to get them out again. We invited residents of Bluff's West 2 and Hillsborough, to attend our first HHHA meeting at Pax Christi, the Catholic Church situated at the top of Homeward Hills Road.

The church was a natural meeting place for us since many HHHA members attended services there on Sunday. In fact, Homeward Hills Road regularly had bumper-to-bumper traffic before and after services. For several years, I worked part-time there as a Religious Education Coordinator for Grades 1 and 2, so I found it easy to ask the pastor, Father Tim Power for help.

Father Tim followed our work closely. He always seemed to be there when I needed a word of encouragement and guidance. After my first speech he wrote, "Wonderful job on your speech about Environmental issues; good for you!" Once after I read him an opinion piece I hoped to have published in the newspaper, he admonished me for attacking BFI and not focusing on the health and safety aspects of a landfill expansion.

After hearing about Father Tim's involvement, Tricky Dicky commented that he was going to report him to the pope because the landfill issue was political, not religious. Of course the pope was leading a political crusade of his own at that time: the overthrow of communism. Had he known about Eden Prairie's environmental battle, I believe he would have been very supportive.

Father Tim was inspirational, and I once asked him at a retreat how he got to be such a polished speaker. I wanted to conquer my fear in this area and thought he might teach me a trick or two.

"Do your homework, Susan. Prepare well in advance. Then practice, practice, practice," he said. Father Tim confessed that in his early days as a priest, he struggled to learn public speaking. "When I first gave homilies, I stood on one foot behind the podium shaking. I was very pleased that the big, baggy vestments hid my body."

Interesting concept, I thought. Tim's persistent efforts to excel eventually produced an orator of unparalleled excellence in the community. Sunday Mass was like a weekly Sermon on the Mount, drawing throngs from all over the metropolitan area to hear his stories. If I ever got the nerve to try, at least I knew where to begin – dressed in oversized clothes, with a well-prepared speech.

September, 1986, Father Tim greeted neighbors as they filed into the Dorothy Day social room at Pax Christi to kick-off the Homeward Hills Homeowner's Association. I lamented to Tim that only 50 people came, but he felt that signified an encouraging beginning.

Ivan chaired the event as we listened, drank wine, and ate cheese. He was our natural choice for moderator since he often gave sales and marketing seminars at work. Later, when I chaired the first dump subcommittee meeting, one member of the group, expecting to work with my charismatic husband, asked, "Where's Ivan? I thought he would be directing this group." I explained that my husband acted as chairperson for the first HHHA meeting because I wasn't ready to assume that role then. However, I was taking over now. He seemed deceived. Ivan and I just laughed. All the same, I worked hard to match my husband's style.

After Ivan discussed the proposed framework for the Association, we voted for an Executive Board. Mary, Tim, and I were chosen to represent the Bluff's West subdivision. The citizens identified four neighborhood problems: school boundaries, traffic control, park upgrade, and the polluting dump. Then we set up committees to handle them. Ten people volunteered for the dump committee. What a bonanza! I was thrilled.

Fifty families registered to join the Homeward Hills Homeowner's Association that first evening. Spirits and camaraderie were high as we emptied several wine bottles to inaugurate the Homeward Hills Homeowners Association. The words of Robert Frost's poem, *Stop by Woods on a Snowy Evening* described this moment perfectly: "I have promises to keep and miles to go before I sleep, and miles to go before I sleep."

Water testing at the park

At about this time, neighborhood mothers requested that the Minnesota Pollution Control Agency test swamp water in Homeward Hills park for pollutants. It occurred to us the leachate plume had possibly reached the bog waters next to swing sets where our children often played. Officials agreed to come out and take samples, but asked to arrive early on the first day of school. No problem, I thought. We were just happy to get this testing done.

As every experienced mother knows, organizing her children on the first day of school creates a maternal frenzy. No one can find shoes; complaints about breakfast cereal fill the kitchen; buses run late, and beds go unmade. And of course, Mom races for a camera to snap a photograph of her well-scrubbed darlings. In short, chaos reigns.

In September of 1986, a truckload of MPCA officials drove up Jack

Pine Trail to our house. At least I was dressed. That's about the only positive thing I can say for the state of myself and my home. Since I was outdoors already, it was too late to run inside and play Mrs. Clean and tidy up the house. I had been seen.

After greeting the men, to my horror, they asked to use the bathroom. I was too embarrassed to request a private reconnaissance mission first to inspect for an unflushed toilet, wet towels on the floor, or tadpoles in the sink. I just pointed the way and prayed to my patron saint for mercy. I got none.

Later, after the men came out, I nonchalantly checked the condition of the room. Just as I suspected. A balled up wash cloth lay in a muddy sink; no doubt Neil bathed his frog there the day before. No problem with towels on the floor, there were none at all. That was probably just as well. I suspected they were used to dry off the frogs anyway. And who knew if the toilet was flushed before the men came in!

I relayed the story to Leslie and he mockingly proclaimed, "Susan, how can I ever call the MPCA again after this bathroom incident. I'm so embarrassed."

It was a pretty morning. The crisp, clear air was about 60 degrees, so I accompanied the men on their water sampling expedition to the park. As they struggled walking through weeds to the water, they picked cattails for my vase. I guess the bathroom couldn't have been that revolting after all; maybe they had boys of their own and understood.

Several weeks later, we got the test results. The water samples showed no contamination. Possibly the plume hadn't reached the area yet. However, we were never convinced the data was valid unless we did the testing ourselves.

Valued political allies

Election Day 1986 fast approached and we needed political advocates for our organization's "no dump expansion" policy. Once again the dump committee put out flyers, done by guess who, to ask neighbors to call mayoral and city council candidates to state HHHA's position. We wanted politicians to understand our refusal to compromise on the dump for dollars.

I took a turn telephoning down the list, too. Patricia Pidcock's name appeared first under council members, so I called her. She listened carefully to my points given in rapid machine gun fire and made very few comments. But then how could she? I gave her no chance. Patricia later confided in me that my persistence tested her patience in those early days. However, I did convince her, a mother of four children, to hop on her horse, grab her lance and ride into battle with us.

Patricia (McDwyer) Pidcock worked tirelessly defending health and safety of the neighbors living near the Flying Cloud Landfill. She influenced other members of the council to do the same. BFI officials were never safe from her. She sought them out in parking lots, at meetings, in court, and anywhere they happened to linger, and berated them for their ignorant stance on the dump issue. Maybe, it was the Irish background, (her father was a foremost legal official in Cavan County, Ireland,) that produced this spunky character. But, regardless, Patricia became a valuable ally. I nicknamed her "Patricia Madame," because most of the time she behaved like a proper lady.

I came to love and admire her as a wife, mother, council member, and friend over the years. She's a rare human being who will stand up and state her convictions, even if they're not politically correct. By her own admission, she said, "I'm a damn fool at times." I disagree. A world populated by more Patricia Pidcocks would make for a safer existence.

Before I continue the saga, I should explain where I got this habit of nicknaming people. It's all in the family. It began with my father. Instead of calling his children their given names, Susie, Johnny, Kenny and Janet, good old Dad thought Sugarbun, Butterball, Booby, and Cooch sounded cuter. You figure it out. As of yet, my father's reasoning escapes me. I tolerated Sugarbun as a child, but let me tell you, during my dating years that name proved extremely embarrassing.

Dreaming up alternate names for key players in the landfill game took the edge off the day-in and day-out frustrations of battling a corporate giant. Our conversations seemed funnier discussing the latest testimony of "Baseball," or "Patricia Madame's" recent run-in with "Tricky Dicky." We kept laughing to keep sane. Maybe that was my father's philosophy, too, raising five children.

Dr. Jean Harris, another council contender, accepted an invitation to attend one of our dump committee meetings to hear first hand our people's concerns. This charismatic African American, formerly a medical doctor and then vice-president for Control Data Corporation, listened attentively to us all. Impressive was an understated description of this woman. Jean asked for a landfill tour, another of my specialties after flyer-making, which I eagerly volunteered to give. In return, I hoped she would help us on health-related issues since, as a doctor, she carried more clout than Susan Varlamoff, former research assistant in Oceanography from the last ice age.

All the political candidates got our message. They publicly voiced their opposition to expanding the Flying Cloud Landfill. However, the incumbent mayor, who had quietly discussed a compensation package with BFI in September of 1985, was under suspicion. He stated along with the others that he supported our position, but a little voice inside us whispered, "If he betrayed us once, why not again?"

November elections returned our former mayor, Patricia Pidcock, Jean Harris, George Bentley, and Dick Anderson for City Council. We knew a close association with this City Council represented our best hope to win our battle against Browning Ferris Industries (BFI). As citizens of several small neighborhoods in Eden Prairie, we felt like biblical Davids striking out pathetically against a multi-billion dollar corporate Goliath. As determined and courageous as we tried to be, we lacked financial resources and political weight the City officials had.

Five years and going strong

Early 1987 marked the fifth anniversary of neighbor John Boyle's audacious announcement that he intended to fight the landfill expansion. At our Environment Committee meeting (we had changed the name of our group from the "dump committee" to sound less radical) we toasted the occasion with champagne. Joining us for the evening was Minnesota House representative Sidney Pauly. She deserved to raise a glass of champagne with us because for years she worked on our landfill issue, first as a councilwoman, now as a state representative. In fact, the landfill expansion was a major concern for most of her political life.

Naturally, I trembled anticipating the arrival of such an important guest and contemplated guzzling my allotment of champagne before she arrived. What would an important person like her think of me as a chair? Had I improved enough since the days of yore to bluff my way though the meeting?

I worried for nothing. Sidney was a gracious woman who calmly assured us we could count on her help. She volunteered to keep the proper government officials informed about landfill expansion developments by mailing them the latest news clippings. Sidney offered to testify for us at meetings and lobby in front of and behind the scenes in the state legislature. She promised to influence events in any way she could, so government officials could no longer support landfill expansion out of ignorance. Like Patricia Madame, she was a mother of four children, and worried that the landfill and its expansion might adversely affect the health of neighborhood children.

Sidney was our angel. I relied on her help for nearly everything. She gave me gentle pats on the back, suggested the right lawyer to represent our group, demonstrated lobbying techniques, provided financial aid, and kept the promises she made at our meeting. For eight and a half years she committed herself to stopping the landfill expansion.

"We're going to win, we're on the side of the angels," she whispered to me once. I believed her. One Christmas I secretly delivered a little white ceramic angel to her doorstep with a note that read, "With you as the angel by our side, we will win."

Soon after our early 1987 Environment Committee meeting, I received a call from the mayor's secretary inviting me to lunch at Flagship, the nearby posh health club where local politicians sometimes entertained. Flabbergasted, I said yes and wondered what I did to deserve this honor.

Naturally, I arrived at Flagship early with my yellow pad of questions and comments for the meeting. However, I felt outclassed by my surroundings where the community's *creme de la creme* wheeled and dealt. With plenty of time to spare, I took a tour of this state-of-the-art club and observed the athletically trim clientele, outfitted in avant garde sportswear, indulging their fitness fantasies.

After my quick tour, the other invited guests began arriving for our luncheon. The mayor, Sidney and Roger Pauly, Dirk DeVries of the Metropolitan Council in our area, and I, sat around a big circular table in an intimate, private room. While we dined on salads and drank Perrier, we discussed the toxic, Flying Cloud Landfill. An easy conversationalist, Dirk was also a wonderful man, eager to work with us. I listened, talked some, but couldn't eat. Not talented enough to chew food and talk simultaneously, I refrained from eating so as not to miss a morsel of thought. My stomach grumbled for lack of food, but my mind benefitted from the fruitful discussion.

The Environmental Committee reciprocated with an invitation to the mayor, city manager, and city attorney to attend our meeting. We wanted more help from the city and specifically asked them what they could do for us. I questioned the mayor about organizing a citywide campaign to close the landfill. He said he felt only those living near the dump cared about the issue. I assured him he was in error. "Maybe not everyone would be willing to walk two-and-one-half miles around the landfill to protest, but everyone I spoke with wants it shut down." His weak answers continued to concern and trouble us. Was he for or against the toxic landfill issue?

City Manager Carl Jullie offered to send school buses over to Pax Christi Catholic Church to transport people downtown for upcoming meetings. City Attorney Roger Pauly asked the group to study the Environmental Impact Statement and provide three or four new facts as reasons to merit a supplemental EIS, because the Metropolitan Council Environment Committee and full Metropolitan Council would vote on this issue in a few months.

Mary Kennedy and I took Roger's advice and sat down once again to draft a letter. This time we wrote to Josephine Nunn, Chairperson of the Environment Resource Committee, stating our four reasons to implement a supplemental EIS. In our best technical talk, we described petroleum contamination in a well outside the leachate plume near our homes, the increasing amounts of contamination at the western boundary of the plume, and the effect of an expansion on our neighborhood.

The second point in particular concerned me. Several monitoring wells located beyond BFI's fence, and within the subdivision, showed low levels of volatile organic compounds. Only one test sample had been taken from each well. John assured me these VOC concentrations should not be a problem. However, I felt certain we lacked sufficient data to prove these levels insignificant. My past work as a research assistant at Columbia University, enabled me to make this small criticism. During my three years of research there, we ran several tests on many sediment samples covering one small area of the ocean floor. When all data was in, our chief scientist, Dr. Pierre Biscaye, would then draw concentration contours.

We argued more sampling must be done over the next year to prove conclusively that contaminant concentrations and the plume size had stabilized. I discussed this problem on the phone with "John Boy" in advance of the meeting. Big mistake. After I entered the council chambers with my dump-busting cohorts, I spied John and sat down beside him. Like old friends, we exchanged pleasantries on our favorite subjects: VOCs, leachate, and monitoring wells, gossip worthy of Eden Prairie's Dump Queen.

During the meeting, the chairperson referred to my letter and asked John why these wells near homes were sampled only once. John Boy stood up and magically produced another set of BFI data he held concealed in his hands while he talked to me. I was livid. Still unable to defend myself publically, I made no comments for the record. However, when John sat down, I reprimanded him in an obnoxiously loud whisper. He gave some feeble excuse about calling my home to inform me about the reports, but got no answer since I had already left to come downtown. "You Benedict Arnold," I fumed.

After hearing all testimony, the Environmental Resource Committee of the Metropolitan Council rescinded their original recommendation for a supplemental Environmental Impact Study (EIS). The full Metropolitan Council followed suit in December of 1986. I felt trouble loomed ahead for the big vote on the expansion next year.

Sandra Garderbring, former Chairperson of the Metropolitan Council, seemed opposed to the expansion, but newly appointed Chair, Steve Keefe, a chemist by profession, barely listened to our arguments. He believed the county required, could provide, and was prepared to promote additional garbage capacity the Flying Cloud Landfill.

The City, angry they didn't win the supplemental Environmental Impact Study (EIS), sued the Metropolitan Council. Again, the court settled the suit in our favor; BFI had to prepare a study on additional environmental issues. Expanding their leaking, toxic dump was now becoming an expensive proposition for Browning Ferris Industries. The company was pouring millions of dollars into their garbage pit only to see it disappear without results. The dump obviously leaked more than leachate.

Interestingly enough, BFI now hired John Boy's boss to work as their consultant. Since she knew all the government regulations and people, she was a wonderful resource for the company. Hiring regulators was common practice for BFI. It began at the top. BFI's CEO, William Ruckelshaus was the former EPA chief. Think of all the clout he still had in Washington among his former colleagues. Think how little clout the average citizen's group had against BFI when they pleaded their case before the highest environmental agency in the land. It wouldn't be long before we ourselves found out what little clout we'd have.

We also learned corporations often beat citizen's groups by "meeting" them to exhaustion. John Boy told me about a situation in which he was involved where citizen outrage and attendance at meetings began strong, then decreased over time. When opposition diminished, the company and government pushed through their proposal. Upon hearing this, I vowed we wouldn't let this happen to us. As Dick Coller would say later, "We need to get mad and stay mad."

Unknown to BFI, a virtual army of dump expansion resisters moved into the neighborhood with the current housing boom. This prolific group of tenacious eco-activists multiplied exponentially over the years as word traveled through city streets about BFI's leaking landfill. Eventually, we became a whole community of Don Quixotes. Though weary, we dragged ourselves and our friends to meetings year after year "to follow that star, no matter how hopeless, no matter how far." We pretended never to tire of the ordeal, though at times continuing our effort seemed ludicrous.

Time Magazine once described people like us as visionaries. "Visionaries are possessed creatures, men and women in the thrall of a belief so powerful that they ignore all else – even reason – to ensure that reality catches up with their dreams." As visionaries, we believed we could shut down a toxic landfill owned and operated by a corporate Goliath, to protect our children and set a precedent for others.

MPCA visits

In the spring of 1987, we boldly invited MPCA officials to our dumpside development, offering them spectacular views of barbed wire fencing, monitoring wells, and an opportunity to savor the stench-filled air. We also intended to reiterate our commitment to shut down the dump, and find out where they stood on the issue. Once again, we prepared to debate the technical demerits of this state-of-the-art atrocity.

The MPCA was the last state governmental hoop in the expansion process. We wanted to influence its officials on our turf well in advance of the decision. I offered to have the upcoming meeting at my home so Ivan could be relieved of babysitting duties. We needed his technical expertise to outwit the MPCA engineers.

Bill Papic, a crack engineer at Honeywell and father of two sons who played with Paul and Neil (we always connected with adults through our children) also joined us this night. He always relished the opportunity to intellectually duel with the experts on the other side. Bill had a wonderful sense of humor; one of the few qualifying prerequisites for entering our group.

As Benjamin Franklin became famous in Philadelphia for his *Little Richard's Almanac*, so Bill Papic gained renown in Eden Prairie for his sage advice in the form of "Billisms." A classic "Billism" was, "If you want to lose weight, eat less." We knew a real winner when we heard one, and eagerly welcomed him on our team.

Another neighbor who came for the dump discussion was Alex Zubar. The poor guy barely unpacked his bags after moving into the development when he discovered that hiring a trash hauler was an unnecessary expense. A quick toss of garbage out his west window, and over yonder fence would land it in the dump. What an economical location for a house!

Eight environmental committee members, plus five MPCA officials, overstuffed our small living room on March 24, 1987. Government officials must have felt safety in numbers since they came in force. In addition, they hauled along all their media equipment, perhaps figuring if they kept talking and showing pictures, we'd shut up.

They began with a slide presentation of BFI's proposed clean-up system for the groundwater contamination. Minnesota state law required BFI to remediate any water pollution caused by their dump before they could expand it. We found it absurd to imagine any system could purify the immense, polluted, underground aquifers. Trying to clean up this water made as much sense as trying to clean up the Atlantic Ocean.

Part of the proposed clean-up system consisted of an air stripping tower BFI expected to install next to the neighborhood. Air stripping is an excellent method for transferring contaminants from water to air. It accomplishes this by pumping water to the top of a tower, exposing it to

76

the air, thereby allowing volatile organic compounds to disperse. "Tricky Dicky" often said air stripping would not create health problems in the neighborhood because "the VOC's disappear."

Somewhere in the dark recesses of my aging brain, I recall memorizing one of the first laws of physics that went like this: "Matter can neither be created nor destroyed."Maybe Dick cut physics class that day, I don't know. But I wasn't looking forward to breathing volatile organic compounds on my morning jog.

Our men argued the facts well this night. They asked every imaginable question about the design of the system, placement of the wells, and so on. Louis kept insisting they raise the height of the stripping towers to disperse contaminants higher over the neighborhood. He didn't want his little girls sick. Mary asked for more sampling in the test wells near our homes and officials agreed this should be done. Tim then demanded to know point blank how the MPCA could shut down the landfill. Ken Podpeskar said, "We can deny the permit."

The next day when I called Ken at his office, he pleaded exhaustion. Apparently, he and his colleagues could barely walk through the MPCA doors this morning after the grilling we gave them the night before. Ken was awfully kind, I felt sorry he had to undergo this punishment. Before long, he quit his government position to study law.

Later that year in June, the public was invited to hear the same presentation about Woodlake's clean-up system that we heard in my living room. Once again, the meeting was held at Pax Christi Catholic Church, so once again we set out to engage in another verbal brawl with the BFI boys and MPCA engineers. Attendance was good and the public was getting smarter.

Someone asked, "How many landfills have been cleaned up with this proposed method?"

"None," came the answer.

Soon after, we learned that no landfill had ever been cleaned up using this system or any other one. The only landfills given a clean bill of health were those in which the contents had been excavated and transferred to another area. Landfill clean-up was new and untried technology, so serving as guinea pigs for this experiment didn't thrill us.

Other meetings followed. At one, we discussed whether the landfill should take fly ash from the incinerator downtown. This ash is often very toxic because it contains heavy metals such as cadmium and lead. We felt that solvents, refrigerants, and pesticides already buried in the landfill supplied enough contamination to the water; we didn't need the addition of heavy metals. Also, open trucks transporting ash through the city, can allow it to be blown across residential neighborhoods. We vetoed the idea, and seeing our fury, BFI's lawyer "Tricky Dicky" killed the proposal.

The pen is mightier than the sword

My dissatisfaction with the landfill permit process began reaching an all time high. Attending meeting after meeting was exasperating. I felt we were riding a stationary bicycle and getting nowhere. We needed a new angle, something to push us forward. But what would it be?

At the Renaissance Festival near Shakopee, Minnesota, I remembered seeing a plaque that read, "The pen is mightier than the sword." So I decided to set aside my lance, grab my pen, and write a piece for the opinion section of the local newspaper. This would give me an opportunity to communicate with all the citizens of Eden Prairie.

My writing was accepted. I began with these words from Ralph Waldo Emerson's poem, *Success....*: "to leave the world a bit better, whether by a healthy child, a garden patch, or a redeemed social condition. To know even one life has breathed easier because you have lived. This is to have succeeded."

In the article I relayed the latest evidence in the landfill case and appealed to all Eden Prairie citizens to become involved with the issue and stay involved until the expansion was defeated and the dump finally shut down. I closed the article with my own words: "For our city, for our health, and for our children, we must succeed."

After being published in the *Eden Prairie News*, I decided to try the big time and submit an opinion piece to the Minneapolis *Star Tribune's* Saturday commentary section. A call from the editor confirmed that my "Landfills Contaminate the Precious Earth" article was accepted. Apparently, the newspaper receives hundreds of pieces each month and only several are chosen weekly for Saturday's opinion page. It was beginner's luck. In my article I warned government officials that "We have a responsibility to future generations not to sicken them with our poor judgment today."

This wouldn't be the last time I vented my frustrations in print.

Throughout the landfill application process, BFI and government officials heard from all of us, including state and local politicians, lawyers, and fuming Eden Prairie citizens. We tried different formats, too. Sometimes we held headline-making demonstrations; other times we bombarded newspapers with letters to the editor and opinion pieces. Then, some of the more creative members of our group designed eye-catching announcements.

Bill Papic and Ivan got the brilliant idea of running advertisements in the paper, offering freebies from a dump expansion. Bill argued that everyone wants something for nothing, so why not enumerate the possibilities if local citizens do nothing to stop the BFI dump's expansion. They came up with a hilarious piece that read:

In Eden Prairie
a proposed expansion may get you
FREE,
Water contamination
Toxic ash
Congestion on Hwy. 169
New and exciting odors
Dumping into the next century

The shock of our message awoke people from their complacency and compelled them to think about the problems of living in a community containing a county dump. Talk around town now buzzed with the issue.

Rest and recovery

As anyone can imagine, crusading sucks enormous resources from crusaders, leaving them feeling empty from time to time. To fill myself up, I read inspirational prayers, poems, and advice from other men and women in history who surmounted incredible odds to demand fairness. Their words helped sustain my courage in days when defeat blocked our every path. I studied the lives of Lincoln, Washington, Gandhi, Jesus Christ, and Mother Teresa, who became giants in society for defending their principles. I learned they, too, were constantly tormented in trying to cut a path through jungles of difficulties and self doubt. Courage didn't come easily to them, either.

The German philosopher Rudolf Hofmann said, "Truth is the fundamental requirement of human life." This must be so, because we felt so outraged at the stupidity of expanding a leaking, toxic, dump near our homes that we became intensely compelled to prevent it. Besides, how could we expect our children to lead lives of integrity if we refused to speak out and defend them? One of our citizen advocates, Louis Affais, often reminded us that presenting our position was easy. "Just stand up and tell the truth," he said.

Summer came and I decided to take a sabbatical from my dump-busting activities to vacation in Europe. Even Jesus Christ occasionally stole away into the desert alone to escape the crowds and pray. So I decided to take a much-needed break from our grinding crusade to recharge my body and soul.

Paul and I flew to Brussels and stayed with Grandmaman (my mother-in-law) for a wonderful three weeks away from the BFI dump. Together with her sister, Tante Ginette, we drove around the French countryside, marveling at the beautiful landscape and five-hundred-year-old stone farmhouses, their window boxes overflowing with rainbow colored flowers. We stopped often along the roadside to *pique-nique* on local breads, cheeses, and wines. *Quelle belle vie!*

We also traveled to the Normandy coast where we saw beaches and cliffs used in the Allied, D-Day landing during World War II. In a nearby museum, I studied how American and English allies planned and launched their secret attack against the Germans at night, while Hitler confidently slept. I'd remember this tactic.

On our return to Brussels, a large brown envelope from Ivan jolted me back to my crusade an ocean away. It contained news clippings describing Eden Prairie's "storm of the century" in late July, 1987. Ivan wrote the rainfall was so heavy that Pierre and Neil swam in the streets. And River Road, located below the landfill, fell into the valley. In fact, two homes built near the bluff's edge collapsed down the slope. This dramatic incident reinforced our argument that the bluff was unstable and couldn't support another 30 more years of garbage dumping. How long would it take before the entire landfill slid into the river, carrying with it decades of toxic garbage and polluting contaminants?"

I returned home with Paul, rested, clear thinking, and rededicated to continuing the quest. It was a good thing too, because the mother of all battles awaited us in the year ahead.

5 | DEFEAT AND DEVASTATION

Barely over jet lag from my European trip, another dump meeting awaited me at Hennepin Technical Center in August, 1987. However, this one was unique. The Metropolitan Council called the hearing to discuss the just completed study on the landfill's pollution, but Sidney Pauly had another agenda. She orchestrated all her rabble-rousing political friends in what was to be the "greatest show on earth" in Eden Prairie that night. The mayor referred to it as "Sidney's Circus."

For the first time, BFI officials just stood on the sidelines dumfounded, squeezed out by this petite, soft-spoken woman. It gave us a unique opportunity to support our cause and have fun without having to prepare a case. We quickly found front row center, ringside seats near Mary and John Kennedy and sat down. Unfortunately, I forgot popcorn.

Sidney took the center under the spotlights first, and in her quiet, sweet voice systematically castigated the integrity of Browning Ferris Industries (BFI). She said they had used aggressive tactics to expand their landfills, frequently violated the law, and were under grand jury investigations across the country. This was the first time we learned about BFI's modus operandi. (Sidney received the information from Security and Exchange Commission reports her stock broker supplied.) She concluded her remarks by saying that granting the expansion would be "shackling our children and grandchildren to pay for our excesses." The crowd wildly applauded its agreement.

Senator Don Storm followed and expanded on Sidney's line of thought with the information she shared with him. A former minister and gifted speaker, Don slowly enumerated BFI's fines and violations one by one in his loud, booming voice: price-fixing in Ohio, 1,000 violations in Louisiana, sued for punitive damages in Vermont, investigated by grand juries across the country, and pushed through an exemption for a Certificate of Need in Minnesota. The people cheered for him also.

Act Three brought us Eden Prairie's former mayor, Wolf Penzel, who is also a chemist and adamant expansion foe. In a mocking tone, he related to the audience how, in 1970, Flying Cloud Landfill was considered "state-of-the-art." Now that it was badly leaking toxic wastes, he declared that using the same description for the expansion didn't give him a warm feeling. Wolf charged "either extraordinary cooperation or collusion" on the part of the Minnesota Pollution Control Agency (MPCA) for granting landfill expansion permits in the past. He suggested that continuation of the policy by various agencies "may ultimately be viewed as criminal."

"Encore!" shouted the fans.

Linda Lehman, the City's newly hired scientific consultant took the stage next. She criticized the "ever changing design" of the landfill expansion and said that "As far as the design goes, we have a moving target." We howled.

This point became the butt of many jokes. Every time BFI issued a report, we grabbed it to see if they had changed the numbers on the vertical and horizontal expansion. We knew that calculating volume was not BFI's forte in their formative years but it appeared this deficiency persisted.

Patricia Madame, not to be outdone, continued the entertainment by reading statements from Councilwoman Jean Harris and farmer Severin Peterson strongly opposing any expansion. Then, Leslie joined the performance, as well as other citizens, demanding the landfill be shut down. The crowd of 60 was in great humor. They cheered every remark aimed at exposing the corporate character of BFI, and called for the company to hit the bricks and get out of town.

Eden Prairie's mayor sat glumly in the gallery making no comments except some snide remarks about "Sidney's dog and pony show."

The show regrettably ended, and, as always, I walked around to talk with friends. At one point, Mary Kennedy introduced me to Grant Merritt, a well-known environmental lawyer, outfitted in a powder blue suit, who attended the evening's spectacle at the invitation of Leslie. I shook his hand and politely said, "Hello."

We had discussed hiring a lawyer to represent the Association, because we believed some members of the city government were sleeping with the enemy. Sidney Pauly suggested Grant, so I quickly invited him to our upcoming Environment Committee meeting to determine whether he could help us.

Grant Merritt, born and raised in northern Minnesota, formerly headed the Minnesota Pollution Control Agency (MPCA) from 1971-1975. During this time, his face became familiar nationally on Walter Cronkite's evening newscasts for his role in prosecuting Reserve Mining Company for polluting Lake Superior with taconite iron ore tailings. Eventually, the state took its case to the U.S. Supreme Court to prohibit Reserve Mining from discharging its industrial wastes into the lake. Since Grant had a successful track record defeating big business and knew the MPCA, we figured he could be the one to spearhead our efforts.

His wife Marilyn called him Don Quixote, and his quest was protecting the environment. Like Roger Pauly, Grant's ancestors were Minnesota pioneers who settled in the north and discovered iron there. Grant passionately loved the land and inherited a home on Isle Royale National Park in Lake Superior. His stocky frame measured 5'11"; he had intense green eyes and a well-exercised mouth that expounded easily on most topics for hours. Grant's manner and voice could be rough and aggressive, but he also possessed the sensitivity to appreciate nature, opera, and poetry.

After a hard morning lobbying at the state legislature, Grant would often cat-nap under the trees on the Capitol lawn – to absorb nature, I suppose. I laughingly scolded him for this practice. "How can politicians take you seriously if you speak to them with grass clippings covering your hair and pin-stripped suit like green confetti." But I suppose they simply saw Grant Merritt as a man so devoted to the land that he took to wearing it from time to time.

BFI's lobbyist, Irv, lamented to me once, "When you hired Merritt, we knew you were serious about the case."

Together, Grant and I became a dynamic duo. Dean Rebuffoni of the Minneapolis *Star Tribune* remarked, "If we could only harness the energy of Grant and Susan, we could light up Minneapolis."

Ironically, Bill Ruckelshaus, Chief Executive Officer of Browning Ferris Industries, now commanding a million dollar salary, faced his old friend Grant Merritt in an environmental battle royale. Ruckelshaus served as chief of the United States EPA during the years Grant lead the Minnesota PCA. Grant worked closely with Bill on the Reserve Mining case and received kudos from Bill for winning it. Grant considered Bill a friend, and had an autographed picture of them together on his office wall. Now that Bill didn't return phone calls to his friend, Grant put the photograph in a closed drawer to make way for dump memorabilia.

Help from Will Collette

During the summer of 1987, I received a news clipping in the mail from my former neighbor, also named Susan, who had relocated to the East Coast. It featured a New Jersey woman who "dumped a dump." I held onto it for months, then finally decided to track down this superwoman via telephone directory assistance, inquiring how she dumped a dump. Very excited to share her story, she explained the details and told me to call Lois Gibbs immediately at Citizens Clearinghouse for Hazardous Waste (CCHW). Lois Gibbs was the young mother and leader of the citizens' group that convinced President Carter to move their families away from Love Canal. After she won the case, she formed a clearinghouse of information to help citizens direct campaigns against pollution in their communities.

My heart pounded with anticipation when I heard about this organization. I immediately called and spoke with Will Colette, the planning director. In his clear, commanding voice, he advised me to send a letter outlining the framework of our organization and a detailed analysis of the fight. I quickly complied. Will wrote back, "I see several good signs: a good organized group, a willingness to protest, an understanding of the enemy, knowledge of what sleazes they are, and solid support from local government officials."

Week after week, I telephoned Will Colette for strategy tips and morale boosting, and week after week he provided ideas, sent information, and pulled my spirits out of the emotional gutter. A self-described Norman Vincent Peale to environmental activists. Will convinced me we would win. I carried his words of hope in my heart. "Hit 'em high, hit 'em low," and, "Susan, you can do it."

His strong, steady voice over the phone had brought victory to countless citizen groups and I'm sure saved many lives. Will's value to society was immeasurable; his help in Eden Prairie eventually made the difference between failure and success. He shaped a movement that forced government to act on many environmental problems facing the country. Thank God we share the earth with such a man. He's a beacon of hope in a society seemingly bent on self-destruction.

Hiring a lawyer

Our anti-dump expansion campaign now sent tremors throughout the business community, too. Local companies realized that conducting business in a city distinguished as the county's garbage can did nothing to enhance their image. In fact, clients began to run out of town. Our organization soon received a $500 contribution from a city businessman to support our work. He stated that he'd donate a similar amount each month, provided we keep his identity confidential. No problem; we could oblige his wishes. We desperately needed funds to hire Grant.

Unfortunately, Leslie Davis of Earthprotector also demanded this money as payment for his services to us. We felt we had been generous enough to Earthprotector already. HHHA donated funds to help him, neighbors made personal contributions, and even Pax Christi Church provided financial support to his organization. We believed hiring a lawyer would maximize our chances of success and Grant's fees were fairly expensive.

Leslie went into a rage when hearing we decided to direct this business contribution to Grant. He demanded $20,000 and full control over Grant to continue working with us. I pedalled back and forth on my bike between committee members' houses seeking advice. Unanimously, we agreed we could not meet Leslie's demands for more money and power. I pleaded with him three times to reconsider his demands. Finally, he said "I have no more time for you, Susan."

I felt terrible. Leslie Davis stood by us in the early days when no one else had, sponsored the "Ring around the Dump," helped us navigate a path through the tangled permit process, and acted as our spokesperson at many meetings. I had no choice but to send Leslie a letter saying that we could not meet his demands and therefore must discontinue our association with him. We thanked him for his help and wished him good luck in his business.

At first, Leslie's leaving the group hit me like a bomb. He'd been a mentor of sorts. I didn't know if I could continue fighting the cause without him, so I cried. I felt so alone standing before the mountain with no one to guide us away from the sheer cliffs ahead. None of us knew the ropes well enough to lead the way. But there was still Grant.

After receiving our big donation check, I put on my mauve corduroy jumper and pink blouse, and headed downtown to hire Grant Merritt. Grant had his office in the IDS tower, one of the poshest and tallest buildings in downtown Minneapolis, which rents office space to the Who's Who in the Twin Cities' business community. I was impressed.

Doormen dressed in fur coats and top hats stood ready to open the great, glass lobby doors for me. I entered and rode the elevator up to the 43rd floor to find Grant's office. Once inside, I settled myself into a leather sofa placed on oriental carpets and stared at the magnificent grandfather clock in front of me. The ambience felt lawyerly, and I felt nervous again.

What did I know about lawyers? Nothing, absolutely nothing. My only experience involved drawing up a will with my father's lawyer. Was I in over my head again? Probably. The receptionist interrupted my reflections of self-doubt and summoned me into the inner sanctum.

Grant greeted me with a smile and handshake then proceeded to discuss everything but the landfill case: his paintings, his father, and Reserve Mining. Time was money. I cringed to think what this verbal dalliance would cost but lacked the nerve to end it. Finally, after a 45-minute conversation, I signed the papers. Though I tried to act knowledgeable, lawyer vocabulary seemed like a foreign language to me. I had just mastered landfill lingo and now had another set of jargon to learn. Ugh! We agreed I'd give Grant a walking tour around the dump while I explained the background of the case.

Paul fell ill the day of our landfill hike. Thank goodness for wonderful neighbors. Bill and Bonnie Swaim, next door, filled in, often caring for my children when meetings beckoned me. Paul, along with Santa Bear, rested on the Swaim's sofa so I could give Grant a crash course on the Flying Cloud Landfill.

This September morning reminded me of the splendid day in June when we marched around the landfill for the Ring Around the Dump. The air smelled the same and the azure sky was as clear. However, today, as we approached fall, the shadows were longer.

Starting out at 12209 Jack Pine Trail, Grant and I strode up the block to the top of the hill where I pointed out the monitoring wells, located on undeveloped property inside the subdivision. "Right between this set of monitoring wells is the limit of the underground plume," I said.

"No one wanted to buy homes this close to the dump," I told Grant, "so the developer sold the land to BFI. The one house built in this section didn't sell for two years so BFI put one of its own people in it."

"Children often bike on these streets, too," I added.

Walking a half block further in this section put us directly in front of a sign hung on a barbed wire enclosure; that read:

DANGER, KEEP OUT

"I never understood why BFI put this warning sign here if its dump was safe to expand," I exclaimed aloud. "Who is BFI trying to keep out and why?" The mystery pit held many secrets we needed to learn, I mused.

We now began a brisk walk around the circumference of the landfill, beginning along the bluff side. The day afforded us a breathtaking view of the river valley, and we paused to admire it. As I explained to Grant the fragile nature of the wildlife refuge below the bluff, a truck inside BFI's fence zoomed over to watch us. It made me shiver.

"We could be targets for a sharpshooter," commented Grant. We laughed nervously, but carefully watched the vehicle as we cautiously continued our tour. I gave an overview of the history, geology, and recent water and soil test results of the landfill. I discussed the explosive nature of methane and the danger of the adjacent gas pipeline. Lastly, I relayed to Grant our concerns that this "sanitary landfill" might contain illegally disposed hazardous waste. We agreed to investigate this possibility.

Grant conceded that living adjacent to this toxic time bomb, operated by a company with questionable ethics, did nothing to promote sound sleep. It must be shut down. He agreed to take on the challenge.

Shortly after the tour, Grant and I investigated the public record of BFI with material supplied by Will Collette. We concluded that their corporate character ranked well below their mathematical prowess. From what we read in this material we became seriously worried about threats to our lives and our families' lives. Grant and I shared these fears one day in his office. I decided to research the issue more and called a friend whose husband is also a lawyer to see if she could confirm or deny this information. In a quiet voice, barely audible, she said her husband suspected BFI had organized crime connections. Will Collette's material provided strong evidence to support this theory.

When in doubt, I called Will. He talked to me at length about the possibility of BFI hurting me or my family. I explained to Will that Karen Silkwood, the dead environmental activist who blew the whistle on a nuclear facility, wasn't my role model. My young children needed me for quite a few years yet. Will said he felt the risk was minimal, since I acted as part of a large, well known organization. If anything happened to me, BFI would be implicated. Subsequently, the company's reputation would suffer and its profits tumble. Corporate chiefs know how stockholders feel about losing money, so they'd avoid this course of action.

Will said Karen Silkwood acted alone, away from the media spotlight. He suspected her superiors figured no one would link her disappearance with whistle blowing. Will cautioned me to keep visible and make sure other members of the organization acted as spokespeople too. This would provide BFI with a moving target, hopefully confuse them, and take the heat off me. I followed his advice precisely.

Grant and I briefly discussed telephone conversations involving sensitive material. Ivan commented that it was too easy to listen in on a cordless phone such as the one we owned. "If we absolutely must exchange important information over the phone lines," he said, "speak on the corded telephone." Grant added that we must meet privately with the most delicate material.

There was also a question of personal liability. Ivan and Grant talked about setting up HHHA as a nonprofit corporation to avoid bankrupting the Varlamoffs with a law suit. A neighborhood lawyer donated his services to make this change. Again, foresight on Ivan's part saved us from possible trouble and financial ruin.

Fund raising

September, 1987, we convened our annual HHHA meeting again at Pax Christi Church. The crowd was bigger, because at Mary Kennedy's suggestion, we invited two more subdivisions to join our association. Grant was our guest speaker. He told the people that citizen crusades are "won in the arena of public opinion," and concluded his remarks with a thought provoking poem written by his uncle.

> *O mighty sea thy message, in clanging tones is cast,*
> *In God's great plan of progress, it matters not at last;*
> *How wide the reef of evil, how dark the shoal of sin.*
> *The wave may be defeated, but the tide is sure to win.*

We understood we had to keep up the pressure, keep up the pounding until we broke BFI's will to fight back.

Grant was a professional at this kind of fight, but he didn't come cheap. Our $500 monthly pledge couldn't cover his bill, so fund raising became our next preoccupation. Louis and I knocked on neighbor's doors, gave our spiel, and asked for donations to offset expenses which included lawyer fees. People were exceptionally receptive and nearly 100% wrote out checks on the spot. Our first fund raising campaign netted several hundred dollars. Again, we applauded our first effort at this new skill. We were good at patting each other on the backs because we so needed the encouragement to continue.

Even with this extra money, we still lacked funds to pay Grant's bills. In desperation, I called Sidney and screamed "HELP!" She gave me her precious fund raising list. Very nervously, I telephoned everyone, identified myself and my cause, and begged for money. Pride was secondary. Every evening after the children went to bed, I dialed for dollars and despised every minute of it. But at least we received money.

However, we never had enough money, so I continued alone,

knocking on doors every Sunday afternoon. In the bitter, cold winter, I hoped my neighbors would invite me into their homes as I shivered outside. When they did, I presented my story to the unsuspecting listener and asked for a donation. My success rate was nearly 100%. When I eventually hung up my crown as "Dump Queen," I knew I could launch a career as a door-to-door encyclopedia saleswoman.

Mary Kennedy organized a very successful fund raiser with a local businessman who promised to donate $10.00 to the association for everyone who used his company's services. Our bank account grew by several thousand dollars during this campaign.

Then there was the infamous bake sale. Mary and I asked neighborhood moms to donate home-baked goods to sell at an outdoor community function in the dead of winter. Dumb idea! Moms took turns freezing their buns off selling cookies, cupcakes, and bars. For four hours of work, we netted a measly $125. That was the final curtain for HHHA bake sales.

With the quest occupying more of my time, I decided to quit my part-time job at Pax Christi Church Church and devote myself full-time to dump activities. The big vote to recommend or deny BFI's landfill expansion permit was coming up before the Metropolitan Council in the fall. Therefore, I had to work closely with my colleagues to prepare for the hearing on November 18, which also happened to be my birthday.

Prior to the full Metropolitan Council vote, the Environmental Resource Committee (ERC) of this agency would take testimony from both sides and make a recommendation. We expected to present solid evidence to prove the Flying Cloud Landfill expansion wasn't the answer to the Twin Cities' solid waste problem. We also needed a good turnout from the community to convince officials we had broad support. With one mighty push, we hoped to defeat the issue here and now, and go home and live happily ever after near the shut down dump. THE END.

Wrong! Wrong! Wrong! Legions of dragons would nearly eat us alive before we could drive the giant out of our backyards. Unfortunately, this story wouldn't be a wonderful little fairy tale with a fast happy ending, but an epic saga that kept dragging on and on and on.

Preparation for the showdown

The HHHA Environmental Committee worked overtime preparing

for November 18. A newsletter went out through the neighborhood urging residents to attend this meeting. Executive Board Members went canvassing door to door in their neighborhoods to rally support. Then, several days before the meeting, volunteers called names in the Eden Prairie phone book reminding them to join us downtown. Sound like overkill? No. We needed people, many of them.

Jerri Coller, an executive board member of the newly invited Deerfield subdivision, did an outstanding job organizing her neighbors. Jerri was the energetic young mother of two children. She walked the same jogging path that I did every morning at 6 a.m. As Ivan and I dragged ourselves up Homeward Hills Road half awake, she would stride by briskly with her head high and arms swinging, a smile beaming on her fully made-up face. She was *returning* from her two mile walk, and as we crossed paths, we always exchanged a cheery "good morning" and discussed the latest dump news.

"Boy, she's spunky," Ivan once remarked. "I wonder what time she gets up in the morning."

"Too early for me," I grumbled.

Jerri didn't realize it then, but she was training for her reign as the next Eden Prairie Dump Queen.

Sidney volunteered to draw up petitions to be signed by thousands of citizens from all over Eden Prairie, and presented at the Metropolitan Council meeting. We wanted to show we had city-wide opposition to BFI's dump expansion. Forty-five people volunteered to canvas the community.

Steve Frick, a very concerned citizen living several miles north of the dump, tirelessly walked the streets in his neighborhood and stood outside his church, collecting hundreds of signatures. He remained a loyal landfill expansion foe throughout the years. His dedicated effort encouraged us all to keep going.

The teacher's union voted to endorse the shutdown of the dump and urged all its members to sign the petition. The petition traveled throughout all the schools, kindergarten through high school, and finally returned to me with 500 names. Community spirit still thrived in Eden Prairie. It pleased me to know my children were educated in a school system where teachers cared enough to become involved on an issue regarding their health and safety.

We requested editors of both local newspapers to announce the upcoming meeting. They took it one step farther and did feature stories on me. This publicity gave us a wonderful opportunity to get our message in print before the big vote. Seeing my picture on the cover of both newspapers was "awesome" to quote my children. However, there was little time for vanity. The upcoming meeting drew near, so we forged ahead with our plans.

After this publicity, other citizens outside the Homeward Hills area called to offer help. We sent them petitions, information flyers about the dump and told them to go for it.

At about this time, I received a phone call from a Mrs. Schmidt who said she lived in a trailer next to the landfill. She explained that BFI fired her husband Bill from his job at the landfill, and she was ill from what she thought was gas poisoning. She wanted to collaborate with us. I agreed to go right over and hear her story, but was somewhat afraid this rendezvous might be a trap. I explained to my boys where I'd be. "If I don't return, call Mary Kennedy." I told them.

I drove to Flying Cloud Drive, turned left, drove past the dump gates, and saw the trailer right up against BFI's wooden fence. I slowed down, made another left off the highway and pulled up in front of it. A woman peered out of her windows as I cautiously made my way up the dirt path to the door. After I knocked, a slim woman of about 55 years opened the door and greeted me looking like a flash from the past, the 1960's to be exact, and introduced herself as Dorrie.

She had long, blonde teased hair, wore false eye lashes over well made-up eyes, and dressed in hot pants and a knit top. Wow! What a figure! I should look so good at her age. She reminded me of an aging Dolly Parton. After I got her dump story, I would have to get her exercise program.

In a whisper of a voice she presented her husband Bill, and asked me to sit down at her small kitchen table attached to the trailer wall. Bill began explaining to me that he had worked at the landfill from 1973 until he was let go in 1984. Evidently, he was displeased with the arrangement so he began a narration of whistle-blowing anecdotes. Trucks came in at 4 a.m. one morning to dump loads of hair spray cans into the dump, violating regulations. Then, on another occasion, partially filled pesticide drums were buried there. He showed me a photo of hazardous waste

92

barrels that sat outside for years before they were properly removed.

Did I know that the current landfill manager threw out all the office files when he took the position in the early 1980's? Back in the 1970's, Bill said, BFI managers ran the operation seven days a week, 24 hours a day in violation of permit regulations. He also detailed many fires at the facility, saying most of them were never reported. Landfill employees just put them out with hoses. I took copious notes as fast as he spoke. I couldn't believe my good luck!

Mrs. Schmidt added that her daughter, Dawn, worked at the landfill for several years before she was forced to quit due to sickness. Her doctor felt the dump environment caused her illness. The Schmidt's and I secretly decided Dawn must write a letter that we would read at the Metropolitan Council meeting.

Dorrie also showed me her diary, where she had reported the state of her health each day and logged in BFI activities. I quickly went through it making notations about numbness, dizziness, nausea and smells of putrid air. The time was getting late and I had to return home to my children, but I promised to stay in touch. Bill and Dorrie, in turn, volunteered to spy on BFI and keep me abreast of their current antics. I felt like I won the lottery with this encounter.

However, the Schmidt's disclosures depressed me, too. If Mrs. Schmidt became ill living next to the dump, then so could we. Bill also confirmed our worst fears about hazardous waste in the landfill. He explained that most likely there was much more; he didn't know everything that was dumped in the landfill during the night. And all the additional fires added to my worries. This was a nightmare. We had to shut down this landfill.

Quickly I telephoned Grant and Sidney to give them the news. They were both excited and thought we might have evidence to clinch the case. Sidney even lost a night of sleep over the possibility. Grant proposed to call Dawn and arrange for her to write a letter.

Public speaking debut

Before our ERC meeting convened for the expansion vote, its members met October 6th to assess the results of Flying Cloud's latest pollution studies. Normally, Leslie acted as spokesperson, but since he no longer worked with us, I decided to try to fill the role.

Simply put, public speaking terrorized me. I'd have preferred lying naked on a bed of nails or walking barefoot on hot coals than standing up and addressing an audience. But I was adamant about getting our points across to government officials, so I decided to use this October 6th session for a practice run as spokesperson. If I failed miserably, I wouldn't do much damage at this minor event and could slither away like a snake without many people knowing I even tried. If I succeeded, I'd gain the courage to testify for the important vote and possibly make an impact where it counted.

I wrote and rewrote my speech, then practiced and re-practiced it for two weeks ahead of the meeting. The evening of October 5th, I coerced my family into critiquing my presentation. I pulled some kitchen chairs into the living room and everyone sat down as I stood before them, nervously giggling my way through a dress rehearsal. After some stern remarks from everyone, I got serious and performed better. Carol Culp and Mary Cooper also called to wish me luck and offer prayers for my success.

Thereafter, I came to rely on Mary Cooper's prayers for each major speech. I'd call her and ask her to mark her calendar with the time and date of my presentation so she knew when to send her words heavenwards. Mary was very religious. I always felt she had a direct line to God, thereby increasing the probability He'd hear my plea for divine guidance. It worked every time.

I also asked Ivan to accompany me for moral support on October 6th. He agreed and took that afternoon off from work. I arrived early, dressed in my yellow and white oversized sweater which I wore over baggy yellow corduroy pants a la Father Tim. Having time to spare, I paced nervously outside the meeting room, waiting for Ivan and Grant to arrive.

As I maniacally walked back and forth, I nearly slammed into Grant coming off the elevator. He appeared startled, first by my sudden appearance two inches from his face, then by my blinding yellow attire. It took him a few seconds to recover. Then, with a half smile, he mumbled something about "nice yellow outfit, Susan," The way I was flying back and forth in the hallways, he probably thought he met up with an overgrown canary.

Ivan came in soon after, and I slouched down in the seat under his arm as the meeting began. Ivan reprimanded me for acting like a wimp (very accurate) and advised me to sit dignified next to my lawyer. Reluctantly, I moved up one row and sat properly beside Grant.

Of course everyone testified before me, so I had to sit nervously for an hour until my turn came. When the chairperson called my name, I hesitantly walked the few steps to the microphone, placed my well-worn speech on the podium, clenched my hands behind my back and began speaking. To my utter amazement, my words came out slowly and measured.

"We have analyzed this landfill bit by bit for years," I said, "perhaps losing sight of the bigger picture, one that I know well and never forget, living two blocks from it.

"The Flying Cloud Landfill leaks toxic compounds into our drinking water aquifers, qualifies for the state Superfund list, has a leachate plume that creeps closer and closer to our homes every day, and is surrounded by 700-800 young families. We worry every day about a Love Canal in our neighborhood. For us the thought is a very scary possibility.

"Browning Ferris Industries has worn out its welcome in Eden Prairie with broken promises, manipulated information, and withheld data."

Lastly, I tried expressing my sincerity of purpose. "I am not seeking public office and am not paid for this work. But as a biologist, I cannot permit our earth to be contaminated without speaking up. As a believer in the truth, I cannot permit Browning Ferris Industries to continue its business of deception in my neighborhood. And lastly, as a wife, Mother, and chairwoman of my neighborhood group, I seek to protect the health and well-being of my family and my neighbors' families."

I ended with a quote from Chief Seattle and sat down exhausted. Both Grant and Ivan congratulated me on a good effort. Dottie Rietow, a committee member, whispered "good job" in my ear after the meeting ended. I felt confident enough now to attempt the November 18th showdown.

Preparations for the meeting continued. Needing all the support we could get, I called all the mainstream environmental groups for help. I wasted my time. None of them would touch the issue. Jodi Thomas was correct in her assessment of these organizations at Louis' home. They all said our cause was too political. I began getting angry with these people and asked them point blank what purpose their group served if not to prevent destruction of the country's largest urban wildlife refuge. They didn't seem to care. No argument I presented would convince them to speak out on our behalf. They wouldn't get a dime of my money in the future, I vowed. I later learned they would never need my money. Browning Ferris Industries and Waste Management financially supported many of these groups. No wonder they wouldn't touch the issue!

Someone had given me names of several scientists at the University who might testify for us, so I tried that avenue. Another impasse. The scientists I spoke with felt no one from the government would listen to them. They complained they had tried in the past to speak out on certain issues and were just ignored. The only advice they offered me was to continue to rally the people and stay the course. "This is your best hope to persuade the powers-that-be to stop the landfill expansion."

It depressed me to hear that the government disregarded the advice of our most brilliant minds. The sadness in their voices hurt me terribly. Why were opinions from these learned men tossed aside like trash while corporate officials from BFI managed to have the government's undivided attention when promoting their business interests? I wondered if, as a society, we weren't collectively losing our will to defend honor and justice.

As days grew closer to the Metropolitan Council meeting, I spoke on the phone nonstop making final arrangements, getting out press releases, raising funds, rallying the troops and so on. The stress of the situation and the strain on my voice probably lowered my resistance to infection. As a result, I became ill with the winter's Far Eastern Flu and my voice turned hoarse.

My shoulders sagged under the weight of representing the citizens as their spokesperson. A poor presentation could ruin our chances of success. I had to be good. But my experience was so limited. It was difficult to rest, and more impossible to sleep well. My sickness hung on for weeks.

During this time, my mind wandered to the bible story where Jesus' disciples discussed how they felt being chosen to preach the gospel to all nations. We're just fisherman, they protested. Teaching is not our profession. You got the wrong guys! "Peace be with you," Jesus said. He explained he'd chosen the least among them to teach the good news about the Savior to all mankind.

I found this story somewhat analogous to our situation. Here I was a homemaker and mother without much expertise in anything but diaper-changing and pie-making, trying to raise the level of consciousness in society to protect the environment. The task overwhelmed me, but I felt I must try. If a few fisherman succeeded 2000 years ago in preaching the good news of redemption, then why couldn't a few mothers influence government officials to change their thinking today on solid waste management?

I worked long, long hours perfecting my speech. Mrs. Schmidt gave me her slides of hazardous waste barrels from the landfill, and her daughter Dawn wrote a good letter, both of which I incorporated into my presentation.

Grant Merritt, Sidney Pauly, Don Storm, and Hennepin County Commissioner Randy Johnson had also agreed to speak. In addition, City Attorney Rick Rosow, Mayor Gary Peterson, Linda Lehman, and lastly, Patricia Madame agreed to testify.

City Manager Carl Jullie kept his promise to send school buses out to Pax Christi Church to bring citizens downtown. We lobbied ERC members on the phone and gathered thousands of signatures on the petitions. We made signs and banners. We took children out of school to show them government in action, and Father Tim prayed with his colleagues for our success. We had businessmen, pregnant moms, children, the torch of truth, and the angels on our side, how could we lose?

Hours before the meeting,, ERC member Dottie Rietow hurriedly telephoned Sidney to inform her, "You don't have a prayer to win the vote." She said Steve Keefe, Council Chair, had made up his mind Flying Cloud Landfill should be the county dump and felt our environmental considerations were negligible. Since there was little landfill space left in the Twin Cities to accommodate burgeoning trash, and not enough time to site another facility, he believed an expansion of Flying Cloud Landfill was the path of least resistance.

Sidney quickly telephoned me with this news. I felt awful. However, being an optimist, I still believed our testimony could persuade Council members to change their opinions. After all, Steve Keefe was a chemist. When we exposed BFI's record of illegal dumping, and showed evidence the landfill polluted water, air, and soil near our heavily populated neighborhoods, he would be persuaded to deny the permit, wouldn't he? Such was the logic of an incurable idealist named Susan Varlamoff, November 17, 1987.

The Metropolitan Council votes on the expansion

After a restless night's sleep, I mentally prepared myself for the day. As I look back, I have difficulty remembering the events. Only a blur remains, which is probably just as well, for it lessens the pain. I remember dressing in my full brown skirt, orange sweater with matching turtle neck, and long silk scarf. Back and forth over the carpet I paced, practicing my speech in a hoarse voice.

Several hours before I made the trip downtown, I walked up to Pax Christi Church and went around the St. Francis Sanctuary set in the woods, to read The Canticle of The Sun. The words are engraved on wooded plaques, spaced at intervals along a trail that winds its way through trees and a wildlife preserve. Before most major speeches, I'd come here to quietly reflect on the task before me. Being more sinner than saint, prayers helped keep my equilibrium by providing the strength and wisdom I sorely lacked. Here was a sampling:

> *All praise be yours, my Lord,*
> *through all that you have made,*
> *And first my lord Brother Sun,*
> *Who brings the day;*
> *and light you give to us through him.*
> *How beautiful is he, how radiant in all his splendor!*
> *Of you, Most High, he bears the likeness.*
> *All praise be yours, my Lord,*
> *through Sister Earth,*
> *our Mother,*
> *Who feeds us in her sovereignty and produces*
> *Various fruits and colored flowers and herbs.*

On returning home, I drove my boys to Pax Christi to take the school bus downtown, then left immediately for Minneapolis, alone. I needed this quiet time to go over my speech one last time before the fireworks began. I also wanted to arrive early at the council chambers to save front row seats for the anti-expansion group.

As I waited with Grant Merritt in the lobby of the Metropolitan Council building, a crowd of laughing people poured through the doors. I recognized them: Mrs. Schmidt, Pete Sadowski, Jerri Coller and other friends from Eden Prairie. There were hundreds of them! Thrilled with the incredible turnout, I punched Grant's arm. Poor man, he smiled his weird half smile (the same one he smiled when I knocked into him coming off the elevator at the last meeting,) spun on his heels, and headed with the crowd up the elevators to the third floor. Grant probably figured he'd get injured if he stayed around me much longer.

The dump buster regiment monopolized all the front row seats in the room. We tried to control meeting rooms by occupying these positions. When someone addresses an audience, their eyes naturally catch the people in the front rows. If those persons supported us, we'd greet them with smiles, laughs, cheers and other good vibrations; if they happened to represent the opposition, we'd scowl, moan, groan, cross our arms and try unnerving them. It was a technique that worked well.

The children, all twenty five of them, including my son Paul, spread out in the aisles to draw pictures and color and eat popcorn to pass the time.

The press arrived in droves. Reporters from four newspapers sat at the press table and pulled out their pads, while four TV camera crews set up tripods to record the action.

I requested that citizens speak first, because we were tired of second class status at meetings. Besides, it was my birthday and I deserved special treatment. I got my wish. However, we should have been suspicious of the Metropolitan Council's sudden good will.

Sidney, Don, and Randy Johnson, our Hennepin County Commissioner, preceded me and spoke eloquently on our behalf. Sidney reminded the ERC that "Future generations are going to have to live with the one vote you make today."

While others made presentations, I stood up and paced back and forth on the side watching and listening. I remembered Dirk DeVries saying this room was constructed to intimidate the people testifying. At one end of the long, rectangular chamber was the Environmental Resource Committee seated on a foot high, horseshoe-shaped stage.

Facing them on ground level was the podium. The audience sat behind the podium. Trying to appear confident with the entire committee staring down at you under bright lights and your supportive friends behind your back is tough. I understood now why there wasn't a list of volunteers vying for the post of HHHA spokesperson.

Finally, after several hours, I proceeded to the microphone, set my speech on the podium, held my hands behind my back, a repeat performance of the last meeting, as four TV cameras zoomed into my face. Chairwoman Josephine Nunn wished me a happy birthday, and a businessman demanded to know which one. In a very hoarse voice, I replied "29 again," and laughed. Then shifting from one foot to the other, I began slowly and loudly. There was dead silence in the room; even the children sat still.

I began by saying that Eden Prairie citizens from all backgrounds had involved themselves in this effort "to show you we truly care about our environment and don't care to deal with BFI any longer." I referred to BFI as a master manipulator and described the MPCA as negligent in monitoring the company's activities. Again I expressed our concerns that the Flying Cloud landfill might be another Love Canal.

Finally, I thrust forward the *coup de grace,* or so I thought; the slides of hazardous waste and Dawn Schmidt's letter. The crowd gasped when they saw pictures of broken drums spilling their contents onto the ground. Then I read Dawn's words describing the total lack of monitoring control at the landfill gate, including her eye witness account of full 55 gallon drums being smashed open in the dump, followed by a horrible chemical "stink". On one occasion she said, a stillborn baby arrived with garbage on a truck. Again citizens gasped.

"Can you in good conscience allow this to continue?" I demanded. "The path of least resistance makes crooked rivers and crooked men."

The room burst into prolonged applause and I sat down, glad it was over.

Grant Merritt followed me, expounding on the pitiful environmental record of BFI and its criminal convictions. Grant also noted that granting the expansion of this landfill would give BFI a monopoly on the region's waste disposal business.

"Is this the kind of corporate citizen we want to have more permits in this state?" he asked. The crowd roared its approval.

After the dump expansion resisters had their say, our favorite BFI official, lawyer Dick, stepped up to the microphone. Predictably, he grinned, ran his fingers through his multi-colored hair, and told the government, "No additional environmental degradation should result from the disposal of additional waste at the site." People jeered.

"Come on Dick, give it up," I muttered.

City officials followed and our mayor talked tough for a change. He bellowed that, "approving a little contamination is like saying it's all right to have a little crime and it's all right to have a little rape." He added, "We've got a bomb and it's not just threatening to go off, it's going off now."

Linda Lehman remarked, "The underground plume isn't going away. It's only going to get bigger." She added that contamination levels at Flying Cloud were equivalent to those at some infamous hazardous waste sites, namely Chem Dyne of Ohio.

After both sides stated their opinions, the ERC discussed the issue before the vote. A TV cameraman came over to tell me, "Don't worry, you have the vote sown up in your favor." Grant confidently predicted a 4-2 win for us. I sat in the front row next to Grant and Sidney with my heart pounding for hours waiting, waiting, waiting for the outcome.

Steve Keefe, chair of the full council, then asked John his assessment of an expansion at Flying Cloud. "John Boy" responded by saying that the Eden Prairie site meets location, operation, and environmental requirements, and should be expanded. In a few short sentences, Steve Keefe quickly explained that the Twin Cities needed landfill space and our site was fine. End of discussion.

Steve called for a vote. The TV cameras rolled. Four hands went up in favor of the expansion, two opposed it. We lost.

We sat astonished and outright insulted. Numb, I slowly stood up in shock as TV cameras recorded the horror on my face. I stared into space unable to speak. My feet felt glued to the ground. I couldn't move. Color drained from my face. I couldn't believe what I'd just witnessed. Hadn't these officials listened to us? Had they no consciences?

I snapped out of my trance, seeing the Happy Birthday helium balloon my boys let up after the vote. How lucky I was to have such a wonderful family! Today I really needed them.

Then Roger Pauly loudly declared for all ears to hear, "We are bowed but not broken."

As we walked out to the car, Sidney said, "Susan even though we didn't win today, you will never be sorry that you tried your best."

Dirk DeVries cautioned me, "Don't expect to find justice on this side of heaven."

We had one last chance to defeat the expansion. It was before the full Metropolitan Council on December 4th. The city sent school buses out to Pax Christi Church again to take us downtown, and this time I rode along with the others. As I walked up and down the aisle talking with everyone, I noticed two older men sitting together. I introduced myself and they told me they were Jim Brown and Harry Rogers. Now I could place them. We often bought fresh corn and pumpkins from Harry's roadside stand after church service. These men were farmers, descendants of Eden Prairie's founding families. With crops harvested for the year, they had time now to support our efforts to protect their land.

The public wasn't permitted to testify, but Council members could make presentations. Our good friend on the council, Dirk, volunteered to make one last ditch effort to bring his colleagues to their senses.

Dirk DeVries, a tall, good looking man in his 50's, was dressed to the nines for the occasion. Standing before the podium in a grey, three-piece suit and polished shoes, he pinched his cheeks for color, then addressed the Council while cameras rolled.

Point by point, Dirk demonstrated that expanding BFI's Eden Prairie landfill would violate the Metropolitan Council's own Solid Waste Policy Plan. He asked Metropolitan Council members to act courageously and implement their own plan. Dirk was our knight in shining armor for the evening.

Then came Steve Keefe, again, to ram the vote through. He said "this (landfill) is for good for the region." Josephine Nunn mentioned the ERC's recommendation to expand the landfill adheres to the policy plan. After a couple more comments, the vote was taken. We lost bigger this time: 11-4.

The Hennepin County Board followed suit. Commissioner Mark Andrews sponsored a resolution to grant a temporary expansion which passed by a vote of 4-3. Nearby Dakota and Scott County boards unanimously supported this request. Politically, we were dead. BFI lobbyists had run around to the other county boards convincing them they might have a garbage dump in their districts if they didn't support BFI's Flying Cloud expansion. Their strategy was brilliant; ours ran amuck. All hope to defeat the expansion seemed lost.

6 | BLOCKADE

C hristmas was just a few weeks away and our spirits were far from merry. Despite our glum mood, we decided to cool our guns for awhile, enjoy our children, shop, bake cookies, and decorate our Christmas trees for the holidays. One afternoon, as the boys and I were ice skating at Homeward Hills Park, our association typist, Debbie Dean, informed me that the *Eden Prairie News* named me "newsmaker of the year." Initially this information surprised me. However, considering we gave local and state reporters writer's cramp from covering our never-ending saga, it shouldn't have. Mark Weber, the editor, wrote a heartfelt thanks to the opponents of BFI's Flying Cloud Landfill for speaking out against the dump's expansion while the majority of Eden Prairie residents remained silent. This positive reinforcement didn't raise my morale to stratospheric heights, but it did pull it off the ground and onto the curb for a day.

I remember this time principally as a painful period of introspection and questioning. It hurt me, and all of us fighting the landfill expansion, to hold up the obvious truth before government officials and watch it smashed to bits. Intellectual, moral, and common sense screamed out against enlarging a toxic garbage dump that poured contaminants into drinking water aquifers, a river, and an adjacent wildlife refuge — especially a garbage dump that stood on an eroding bluff overlooking the Minnesota River Valley, and abutted a large gas pipeline, a busy airport and heavily populated scenic residential neighborhoods. Only a fool or a junkie would believe otherwise, and we were neither. We were intelligent, caring, sane human beings most of the time, who could clearly see that a dump expansion in our neighborhood endangered all we cherished.

How could our government officials support such obviously criminal stupidity? Whose interests were they serving? BFI's? Their own? The community's? Even if John's solid waste projections did require additional landfill space, why risk the welfare of thousands of Eden Prairie citizens by expanding Flying Cloud? How did this expansion, with all the strikes against it, travel so far up the levels of government bureaucracy? Simple corporate greed and political expediency, we decided, were the reasons.

Browning Ferris Industries used aggressive tactics, as Sidney said, to cut up and cut out the competition around the United States. A 30-year expansion of the Flying Cloud Landfill would give this three billion dollar corporation many millions more revenue over the life of the dump, plus a monopoly on landfill space in the Twin Cities. Once BFI controlled the dumps, they could hike tipping fees through the roof, raise garbage pick-up prices, and increase profits to monumental heights. Money was talking; as a matter of fact, it shouted and screamed.

Politically it seemed easier to expand an existing dump than site a new one requiring ten years and millions of dollars to accomplish. How about a mandatory recycling program to reduce landfill need? The current feeling was that not everyone would participate so it wouldn't be economical on a large scale. Therefore, county planners preferred to bend the rules a little to conduct business as usual.

The cooperative effort of BFI and state politicians made for extremely tough opposition, but I always believed reason would ultimately prevail if we kept up the pressure. But how long could we all hang on? How long would our crusade take? Did our city have sufficient funds to continue the fight year after year? If need be, would citizens support increased taxes to stop the expansion? Were we right to assume this landfill was a health and safety hazard? If BFI got desperate for expansion, could I be a victim of the mob? Time and time again I agonized over these questions, searching for answers.

Joe Mengel, an enlightened scientist, made the observation that if we allowed the landfill to expand at this blatantly unsuitable site, we would set a dangerous precedent for future cases. Then any land would be permissible for garbage dumping. Quickly our rivers, streams, aquifers, and wetlands would become toxic, poisoning life all the way up the food chain.

In acknowledging our oneness with the earth's elements, the cause was raised to a higher level and took on a deeper meaning. We were fighting to set a precedent for future generations of the human family. The buck must stop at our dump.

As I mulled over these ideas, I intuitively felt that expanding the landfill posed a significant risk to adjacent neighbors and the wildlife refuge, but we needed some hard data to support this assumption. If we believed this premise, then it was necessary to act swiftly and decisively to win a battle, because people were losing faith in our ability to be effective, and without their continued support, nothing could be accomplished.

The alternative was to wave the white flag of surrender. However, could we live with ourselves if we didn't try to protect our children. Even if we lost the war after a bitter fight, we could confidently say at day's end, "We courageously defended their well-being with our very best effort."

So what if it took a few years to accomplish the goal. What's a few years in a lifetime of 80 years? Rarely in people's lives do they have the opportunity to alter the course of events for the betterment of humankind. I believed this was one such opportunity.

Life had been generous to our family. Ivan earned a good living, allowing me time to pursue a community project instead of having to work to put food on the table. Most important of all, we were healthy.

These few precious years presented a unique time in my life, where I could throw caution to the wind, sharply focus my vision, and move forward with boundless energy and optimism on a monumental quest. Just a couple years earlier, this would have been impossible, because my babies required constant care. And several years into the future, my teenage boys would never tolerate living under the same roof with Mom Quixote. After that, I needed to work to pay for college tuition, and after launching the boys into the world, I might be too wise and too tired to believe I could overturn the status quo in society. However, as a young mother, inexperienced in the world of politics and business, I decided the angels were going to win this one.

But first we had to develop a battle plan to reverse the government's pro-expansion bias, and we had little time to do it. I brainstormed with Grant, Sidney, and Will for hours on the phone trying to hack a path through this tangled wilderness. We concluded BFI was deeply entrenched in state politics. If we expected a fair hearing, we'd have to force the expansion issue out of the political arena and into the courts before an impartial judge. Lastly, we must publicly show government officials Eden Prairie citizens wouldn't accept being relegated to the role of state garbage can any longer. More and tougher citizen protests were in order.

One item that headed my immediate agenda was to ask the Eden Prairie City Council to pay Grant's legal fees. After our loss at the Metropolitan Council, it became nearly impossible to raise funds. The whole exercise exhausted me, wasting precious time I desperately needed to guide the organization along our currently difficult path. I explained to the city council that so far we raised $20,000 in donations, evidence of our commitment to shut the dump. Would they hire Grant so I could get on with running the association? After considerable discussion, they agreed. What a relief!

After the Metropolitan Council vote, we noticed BFI officials strutting confidently about town as though they'd won the war. In January, at an MPCA meeting downtown, "Tricky Dicky" sneered at me and said, "Susan, the whole government is against you. You don't have a chance." I nearly burst into tears on the spot. I knew he was right. A friend yanked me out of the room before I reduced myself to a blathering idiot in front of this insensitive man. When I regained my composure, I stiffened my resolve to beat those bullies with whatever it took. Damned if they thought they could insult our intelligence, squash us like bugs, and risk our families' well being to fill their pockets. They wouldn't get away with this. Just you wait, I thought to myself.

For a few months we adopted a low profile in the media to let BFI believe they won the case. Meanwhile, we planned our major assault. I remembered the story of the secret allied invasion of Normandy while Hitler confidently slept. It turned the tide of the war. However, unlike the allied invasion of Normandy, we lacked ideas for a plan.

On someone's suggestion I called the I-Team, an investigative reporting team willing to risk their necks for an exciting TV story. Since everyone talked garbage these days, and toxic dumps were big news nationally, we imagined that team might like to do our story. A team of reporters drove out to my home immediately, questioned me for hours about the issue and toured the dump site. However, they said a television story required action shots of BFI in a compromising situation. Right now the landfill remained shut to all but private trucks. What they wanted to show was a good midnight illegal dumping spree. Maybe the Schmidt's could provide us with a lead as they snooped over the BFI fence, I thought. I proposed to stay in touch and put Dorrie on the alert for any suspicious night activity at her neighbor's place.

I pestered these reporters often to do something with this story because we felt a television expose of BFI would shake them into submission. How foolish the MPCA would look if the I-Team caught BFI in illegal activity while government officials insisted that Woodlake's record of violations was minimal. However, the I-Team always seemed occupied with other assignments when I called. Did BFI get to them?

Next, I secretly telephoned the FBI. An interested businessman across town suggested I call the hazardous waste division of the FBI, because it's a federal offense to dump hazardous waste illegally into a landfill. If we could prove this violation, then the federal government could prosecute BFI. So I telephoned them, too. The bureau agreed to send an agent out to the house to investigate. We took no chances discussing this matter on the phone in case my telephone was tapped.

A few days after my call, there was a loud bang on my front door, accompanied by a strong female voice announcing, "Sue Boyle, FBI." I opened the door to find a young, pretty, diminutive woman standing there, a stark contrast to her voice. I invited her in and we discussed the case. She was eager to help. I explained that we knew hazardous waste had entered the landfill since Bill Schmidt, an ex-employee, told us so.

Sue agreed to speak with him.

After meeting with Bill, Sue and I spoke again, and she suggested we use the agency as a last resort to stop the expansion. She believed we had a strong chance to defeat the case before the MPCA. If we didn't prevail, only then would the FBI dig up the landfill, drag out any hazardous waste barrels and prosecute. Since excavating a garbage dump emits a foul odor, she recommended we use this strategy only if all else failed. She said the FBI would carefully watch the case. That made perfect sense. I agreed.

We didn't idly sit waiting for results from any of these efforts, but continued nosing around town, pursuing other courses of action, hoping someone or something would lead us out of our quagmire.

Politically dead

In January, Sidney Pauly drove Corinne Hensley, a new member of our group, and me downtown to the Board of Ethics to look over the U.S. and state politicians' campaign contribution list. How very interesting! BFI was a popular contributor to the campaigns of the Governor, Attorney General, and many other politicians. Their political clout ran deeper than we thought. Over lunch in the legislative cafeteria, we realized that we definitely had to get the landfill expansion issue out of the political arena. BFI was too cozy with too many politicians.

We relayed the news to Grant Merritt and Roger Pauly. Both lawyers decided to request that the MPCA grant a contested case hearing. In a contested case hearing, both sides present their testimony before an impartial judge who evaluates the evidence and makes a recommendation to the MPCA staff. There is no jury. Then the staff considers the judge's recommendation and can accept or reject it. However, if the judge ruled in our favor, we believed the MPCA would have a difficult time rebuffing his opinion. Since we felt there was enough evidence to support our position, we decided to pursue this path.

Around mid-January, our lawyers also sent a letter to the MPCA staff reminding them that the permit expiration date for the Flying Cloud Landfill was January 28, 1988. This date was burned into our brains and we were anxious to give the MPCA the opportunity to act on our behalf. So we waited for their response. None came. Par for the course, we figured. The landfill conducted business as usual. Small pickup trucks

110

entered the gate, deposited their contents, and left. Flying Cloud was also a regional headquarters where BFI stored their dump trucks so there was always plenty of activity.

The process of enlarging the dump continued unhindered.

Since the Metropolitan Council recommended the landfill's expansion, a permit passed to the hands of the Pollution Control Agency staff, where it had to be decided whether BFI's Flying Cloud facility met state pollution control regulations. But this wasn't the final decision. A Board of citizens appointed by the Governor had the last word and could overrule the MPCA. Not wanting to leave matters to fate, BFI's lawyer Dick and his cronies moseyed on down to the MPCA to banter with the staff and perhaps invite them out to lunch to extol the technical merits of BFI's expansion and the Twin Cities dire need for additional landfill space.

Unfortunately, MPCA's staff consisted of inexperienced young people who were overworked for pitiful pay. They could easily become tender vittles for the voracious appetites of BFI's well-seasoned lobbyists. Day in and day out dealing with the enormous strain of angry citizens calling in, plus government officials pressuring the agency to expand the landfill, BFI lawyers perpetually swarming the corridors, and the media persistently probing the staff for comments, could drive any reasonably stable individuals whacko.

We designated MPCA call days where Eden Prairie citizens jammed the agency's phone lines registering complaints to balance BFI's "on the spot" impact. Despite our enormous efforts, Carol, the latest MPCA recruit to our case, recommended the expansion of BFI's Flying Cloud Landfill as environmentally safe.

I suspected this was Carol's first job out of college; she looked all of 22 years old at the time. It was frightening to think the lives of thousands of citizens and the destiny of a huge wildlife refuge hung on a decision made by this very young, inexperienced woman. I believe she, too, was coerced into making this recommendation by the political powers-that-be and BFI. Eventually, Carol fell quite ill and was sent on an extended vacation around the world to recuperate.

The Surgeon General should have stamped the case with "Warning: Hazardous to your health" or with "Danger, Keep Out," which is BFI's own warning posted on its Flying Cloud landfill's barbed wire fence. Our record for driving people off the case was nearly 100%. The dead and wounded soon began piling up. The time was right to ask for a contested case hearing. Grant proposed we bypass MPCA staff members who were bosom buddies with BFI's lobbyists and ready to grant the expansion. Our strategy instead was to make a public appeal to the MPCA Board. This was our last and only shot to get a fair hearing. We felt MPCA Board members, who were more experienced and older, ordinary citizens appointed by the Governor for their work or interest in environmental pollution control, would consider our concerns more carefully and be less likely to fall victim to BFI's slick lobbyists. Since these people lived all over the state, making house calls would be difficult for BFI.

In February the Board proposed to hold a public hearing in Eden Prairie to hear citizen testimony on whether to grant a contested case. Government approval for the expansion permit was nearly complete now. This was our last stand. We had to get local citizens aroused and angry enough to convince the Board we had legitimate complaints warranting a contested case hearing.

Corinne Hensley tediously waded through piles of earlier studies, searching for information and scientists who might convince the Board to grant us the hearing. She learned the U.S. Fish and Wildlife actively opposed the expansion in years past, and one of their former scientists, Dr. Dwain Warner, made a biological inventory of all the plants and animals in the river valley below the landfill. Cory suggested we call him to see what he knew about the situation and ask if he might help us.

Another past opponent of the expansion was Upgrala Management Co., a rifle association that owned the refuge. Chuck Moos was listed in the papers as head of the organization. Cory called both men and hit a bonanza. Dwain agreed to have lunch with Linda Lehman, Cory, and myself to discuss the case. Charles Moos, yet another lawyer, was thrilled to learn that we had taken on the anti-expansion crusade, which he had previously fought until burnout. He promised to back us financially, attend meetings, and act as spokesperson for his organization.

With all the groups and governmental agencies joining Minnesota's environmental battle of the 80's, many lawyers saw outstanding opportunities to make mucho money. Representing the anti-expansion forces were Grant, Roger, his associate Ric, Chuck and two lawyers in

112

HHHA. On "the other side" was BFI's brigade which included Dick, Jerry, Chris, in addition to other lawyers from the Minneapolis firm they employed. From time to time, legal help was also flown in from the company's Houston headquarters, and even Commander in Chief William Ruckelshaus came to town to argue his position before government officials. The Attorney General's Office, MPCA staff, MPCA Board, and the Metropolitan Council also dispatched lawyers to the proceedings.

Finally, the meetings became an orgy of lawyers, whirling around the room in black pinstriped suits and black wing tips, jockeying for the best position to represent their clients, while the clock ticked and the cash register rang up thousands of dollars for them. With the average legal fee at $100 per hour, and an average meeting containing 10 lawyers for three hours, $3000 was billed to the taxpayers per dump seance. And there were hundreds of meetings and court hearings.

Scientific rebels

Lunch with Dwain Warner proved nourishing. It wasn't the delectable meal we ate, or in my case, forgot to eat, but the conversation of this gutsy, intelligent, caring human being that filled us up. Corrie, Linda Lehman and I listened to this 70-year old scientist relate his experiences standing before the Metropolitan Council 20 years earlier, warning officials that siting the landfill on the river bluff, over three drinking water aquifers, would prove costly and dangerous to the state. We sat spellbound ingesting every word.

Next came his show and tell, the monstrous, foot thick volumes 1 and 2 of the biological inventory done for the Minnesota River Valley that he put together at the request of the U.S. Fish and Wildlife Service. We leafed through the pages and saw beautiful pictures of the valley's fauna and flora. Dwain explained this was the largest urban refuge in the entire country. It contained more than 500 plus acres of a flood plain wetland complex that formed an exceptional wildlife marsh. Four hundred blue heron and egret nested there, and many migratory birds used the area. He said the seeps and springs supplied fresh water that fish and waterfowl required for their survival. And these water sources originated from groundwater underneath the landfill. One of the wells in the refuge showed low levels of volatile organic compounds, probably indicating the leading edge of the contaminant plume.

Another of Dwain's concerns was the migration of methane gas into the bluff. Methane replaced oxygen in the soil, he said, and could cause vegetation to die due to oxygen deprivation. If grasses on the bluff died, the slope would become more vulnerable to erosion. Dwain quietly confided that he believed the process was already underway. He was going to check it out by comparing old aerial photos of the river valley with more recent ones.

We begged him to protect his investment of time, energy and love by helping us. He asked for time to give it some thought, but finally decided to join our quest to save the refuge. Dwain asked only that his expenses be paid. We could handle that. We felt so fortunate to have an individual of his high caliber on our team.

Dr. Dwain Warner wasn't only a renowned ornithologist, but a person of incredible cunning. He personally knew several BFI lawyers, especially the young whippersnapper Dick, whom he watched grow up in his neighborhood, and with whom he had worked on previous cases. Dr. Warner knew their intimate legal strategy almost like he knew the inside of his own pocket. He was prepared to beat these "kids" at their own game. No one could outfox this old man, he implied.

Dwain carefully explained to us Dick's, and later Chris's mode of operation to prepare us for their wily tactics. "Watch out for Chris," he said. "He tries to topple the leader expecting others will fall down behind like a set of dominoes."I made a mental note of that information. After this clandestine encounter with Dwain, we felt pumped up and ready to soar again.

Andrew Jackson once said, "One man with courage makes a majority." Dwain is such a man, a truth-seeker who stood up in society, demanding to be heard and willing to take risks associated with his stance. In the end, his prediction for the landfill site, and his understanding of how years of garbage dumping would impact the surrounding environment, hit the bullseye. Again, we found another giant in our society.

Next, we telephoned the U.S. Fish and Wildlife Service, and renewed their interest in actively opposing the expansion. We invited them to come on board, too. Ed Crozier, their chief scientist, offered the group's help with all means available to them. This Washington-based national organization enjoyed enormous respect in Minnesota. Having them champion our cause was a boon.

Our base of support grew now by leaps and bounds; "the more the merrier" was our motto. The more people, and the more organizations that backed us, the happier we became, because we knew our clout rose with each new inductee. We kept our progress quiet, however, since we planned to use this foundation of help for some future action, but what? We still had no clue.

Greenpeace to the rescue

When I read the newspaper headlines one morning, a light bulb clicked on in my head. There on the front page was a large photo of Greenpeace activists hanging precariously from the Minneapolis incinerator stack unfurling banners. That was it! Greenpeace specialized in headline-making activity, and we needed their kind of help. I instantly reached for the phone book, found their number, and telephoned them.

With diarrhea of the mouth, I spewed forth our landfill expansion story, our frustration at losing our first battle, and our hope to stop BFI before we suffered grim and far reaching pollution consequences. I explained that the landfill permit had expired, but the MPCA refused to shut down BFI's operation. The Greenpeace person listened attentively to my plight, then promised to talk over our situation with a supervisor and get back to me. Several days later, Norm of Greenpeace called, listened again to my story and said, "I think we can help you." Since a protest of this size would involve considerable funding, he said he required approval from his supervisor, Ben Gordon, the Greenpeace midwest regional director.

Ben flew out to Minneapolis and furtively drove out to the house to discuss the matter first hand. I was the only HHHA representative present at the meeting. I felt that something of this nature must be carefully explored before presenting it to the whole group. Ben, a tall, thin, intense, and intelligent person of about 30, settled himself with his four colleagues around our circular kitchen table and quickly threw questions at me. I suspected he wanted to know if I understood the scientific and technical aspects of the landfill's problems. Maybe this was a test of sorts.

"Name some of the chemical compounds polluting the water," Ben

115

demanded.

"Benzene, toluene, trichloromethane," I said without a moment's hesitation. I had forced this information into my brain for so many years and was pleased to see I could recall these compounds under duress.

"Explain the landfill's technical make-up."

"It's your basic unlined hole in the ground."

After more questions and answers, Ben asked to tour the landfill site. On seeing the dump's proximity to the neighborhood, he agreed the situation was serious and could only get worse with an expansion. He knew the corporate profile and operating mode of Browning Ferris Industries around the United States and understood why we were getting no satisfaction from the government. Intervention by Greenpeace, Ben believed, could turn the tide. So he agreed to stage a demonstration together with the citizens.

"Timing is everything," he said.

We will schedule this action after the landfill permit's expiration date and before the MPCA public hearing for the contested case.

Ecstatic with Ben's proposal, I presented the results of our meeting to the Homeward Hills Homeowners Association Environmental Committee. At first most everyone was skeptical about a militant protest, but we realized we had few options left with which to fight the neighborhood bully. We had exhausted all conventional means of influencing decision makers. A cooperative militant action with Greenpeace might be our only hope to shock BFI and Government officials into taking us seriously. We reached a consensus to go forward with a Greenpeace demonstration and to keep this activity top secret.

I traveled alone downtown to make plans with the Greenpeace activists at their office. They often worked with CCHW, so sometimes we telephoned Will and spoke in three-way conversations over extension lines. I explained to Will and Ben that my friends and I had no experience in the field of militant protesting. In fact, the whole idea scared us to death. We understood that whatever action we settled on had to be fairly tough to show our resolve, but we lacked the stomachs to chain ourselves to fences or lay down in front of garbage trucks in classic Greenpeace style.

"Please, can you think of something strong but safe?" I timidly asked. "After all, many of us are nice Catholic mothers raising children.

116

And besides, our husbands would kill us if we ended up in jail."

Although a tall order, they sat down to consider the possibilities and came up with the perfect demonstration: a school bus blockade of the landfill entrance. Greenpeace activists suggested that neighborhood mothers and children board the bus, since this would prove our maternal resolve to protect our children from a leaking, toxic dump. What a brilliant idea! And it seemed so innocent!

Our timetable was tight. Only three weeks remained before the MPCA meeting. Ben combed the area searching for a school bus to buy, while Norm and I planned the logistics of the protest. Actually, it was Norm who laid the plan, and I who submissively followed his instructions. I watched with amazement as he and his colleagues set the operation in motion, how expertly they calculated every angle to their advantage.

Norm suggested we schedule the blockade for President's Day, February 12. It was two weeks after the permit expiration date and two days before the MPCA hearing.

"The timing is perfect," he said.

After MPCA board members watched citizens blockade the landfill on the evening news, they might think twice about denying us the right to a contested case hearing. If the officials missed the news, they could read all about it in the newspaper the next day.

Since President's Day occurred on a Monday, a slow news day barring a disaster, we should get good coverage. Norm chose early morning for the protest, because we wanted to hit all the day's news from 10:00 a.m. until 10:00 p.m. Also, an early morning blockade would give United Press and Associated Press wire reporters ample time to write the story and send it out for the following day's papers. And a President's Day blockade would give us the extraordinary opportunity to involve children, since they had the day off from school. These guys were brilliant strategists!

We tried incorporating the flag and a Presidential quote into the scheme. Doug Larsen, our former flag bearer for the "Ring Around The Dump" protest, generously lent us his stars and stripes again. And Ivan came across a Thomas Jefferson quote he remembered from studying to become an American citizen. (He was Belgian.)

117

"When things get so far wrong as to attract
their notice, the citizens, when well
informed, can be relied on to set them right."

These words were ideal. They expressed our sentiments precisely. As spokesperson, still the job no one else wanted, I vowed to repeat Jefferson's words over and over under the American flag at the landfill gate, February 12, 1988. No doubt, this would be one President's Day Eden Prairie would never forget.

Now that we had a plan, we needed people to make it work. For once I took the advice of the sisters at Immaculata College where I studied for my undergraduate degree. "Girls, plan your work, and work your plan," they sang perpetually in our ears as my friends and I pulled all-nighters cramming for exams.

The work was planned, thanks to Greenpeace, and Norm came out to Eden Prairie to gently explain details of the blockade and how it would work. He told everyone up front we would be trespassing on BFI property and breaking the law. "It's the only effective way to close the front entrance of the landfill to incoming garbage trucks," he said. Norm suggested that mothers and children board the bus for reasons he explained to me. He also felt there was less chance the police would arrest us.

The week before the blockade, 12209 Jack Pine Trail bustled with activity. With all this activity, I worried BFI would sense something was happening. However, officials were too confident in their skins to notice anything and relaxed their guard — big mistake on their part. Norm and his colleagues came out to the house to make banners and signs with us. In our basement, the neighborhood children designed a big banner to put over BFI's entrance sign. We all laughed at the result.

Depicted on the canvass was BFI dump headquarters besides a recycling facility. Storm clouds and lightning hung over a small, grey and black BFI building belching smoke. Partially hidden behind clouds was the sun sticking its tongue down at the building. Dead and dying trees, bushes, and flowers surrounded this horrid structure. In contrast, the recycling facility, drawn in bright reds and yellows, stood grandly under a beaming sun. Healthy trees and plants graced the front entrance.

Who ever said, "If you want the truth, ask a child," was very wise.

118

We adults couldn't have matched our children's creativity.

Besides preparing for the upcoming blockade, I spent a good deal of time on public relations. A local businessman offered to design informational flyers for us on his computer. He also convinced a national magazine editor to feature the Eden Prairie fight in the February issue. A color photo of our neighborhood overlaid with a photo of hazardous waste barrels made the front cover.

National Geographic Explorer telephoned with a serious interest in flying to Eden Prairie to photograph the blockade. The editor was assembling a garbage story and thought a citizen protest would visually show the problems caused by landfill pollution. We were thrilled at the idea of hitting the big time. However, I worried about our safety with our blockade and believed *National Geographic's* showing up might complicate a potentially dangerous situation. Later, in a cost-cutting measure, the magazine changed its mind and decided to do a story on New Jersey's great garbage dilemma, which is closer to the magazine's Washington headquarters. I wasn't dissatisfied with the switch.

Dean Rebuffoni also called to do a front page feature story on me. I couldn't believe all the publicity I was generating lately when two years ago my status as a stay-at-home mom ranked me at the bottom of the social strata, probably somewhere between janitor and maid, certainly unworthy of newsprint attention.

In the 1980's I never aspired to be a "Supermom," that incredible female who works by day as a CEO in silk suit, and by night as a mother of three toddlers, and as a wife, gourmet cook, and housewife in blue jeans. After Pierre was born, I hung up my cape and dropped out of the "supermomdom" race because frankly, I just couldn't figure how to accomplish all of it and keep my sanity. Therefore, when my children were young, I made no money, except a small stipend Pax Christi Church paid me for volunteer work. I had no exciting job description and spent my days in total obscurity doing mundane activities such as changing diapers, taking children's temperatures, singing nursery rhymes, baking cookies, and running a religious education program for grades 1 and 2 at Pax Christi Church. All this sudden fame overwhelmed me after ten years of feeling a bit like a social pariah and cocktail party bore.

Early on I realized television and newspapers were necessary tools to win our crusade. The media were like sculptors' hammers and chisels we could carefully use to chip away at public opinion until we created an image we wanted. The way we held these tools was critical. Our every move was scrutinized. Any misstep on my part as spokesperson, like a slip of the chisel, could destroy all our previous work, and only a shattered dream would remain.

"Public sentiment is everything. With public sentiment, nothing can fail, without it, nothing can succeed," said Abraham Lincoln.

History bears this out. Public opinion has defeated armies, toppled governments, and brought down billion dollar companies. What we hoped to accomplish in Eden Prairie, in the Twin Cities, and in the state, was no less than an environmental revolution. We pulled in the welcome mat to a company that showed no respect for our environment, our people, and our law, and we demanded that government develop a strong refuse dump reduction and recycling policy.

During my luncheon with newspaper reporter Dean Rebuffoni, I was gently prodded to discuss my life and current passion. He asked for it, so I'm sure I bowled him over with my nonstop regurgitation of detail upon detail of how terrible the dump was, and how we were going to shut it down and stop BFI's expansion, so help us God. Several times during my monologue, he asked me to stop so he could catch up with his note-taking.

From his days as MPCA head, Grant Merritt knew Dean and described him as an intelligent person with diverse interests, a reporter of integrity, a good father, and a wine connoisseur "par excellence."

As Dean took copious notes on his long, thin, reporter's pad, I couldn't help speculate how he'd feel in our position. He was a father of young children, too. I never found out the answer, but without Dean's interest in our crusade, BFI's landfill might still be open. He provided the avenue through which we vented our disgust and anger at BFI and the government, the latter because instead of protecting the people, they had long since deteriorated to favoring big business.

I confided to Dean we believed hazardous waste had been illegally dumped into the landfill, and could present a real threat to our children's health. I told him "mystery pit" was my pet name for this huge hole in the ground, and that remark made the newspaper. My entire life story came out, too, including where I was born, raised, attended school, worked, all the places Ivan and I had lived, birthplaces of the children, and, heaven forbid, my age for all the world to see.

Dean had no clue when his editors would print the story, but said he'd be in touch to refine the piece. Since he was so kind, I whispered a hot tip, "Be at the landfill gate on President's Day at 9:00 a.m. if you want to scoop the day's headline news. I can't tell you the details, it's top-secret, just show up at the dump." He appreciated the inside information and I appreciated his attention to our cause.

B-day (blockade day) fast approached. I kept badgering WCCO-TV's I-Team to get moving on our story. Then, a few days before our demonstration, the I-Team reporters verbally agreed to televise the landfill story. We figured with publicity like this, sweet victory lay just ahead. Sidney and I had decided to have a champagne and caviar party if the I-Team came through for us, so corks were now popping at the Pauly's home.

All the dump-busting renegades turned out: Grant and his wife, Ivan and I, Don Storm and his wife, and of course Sidney and Roger. We toasted our good fortune at getting the I-Team on our bandwagon, and laughed late into the night. As we prepared to leave, Roger wished me and my maverick friends good luck with the landfill blockade. That brought me back to reality, to the danger lurking a few days away.

Sunday, February 11, was frenetic as I checked final details for the following day's blockade. Greenpeace people confirmed the bus pick-up at 8:45 a.m. in the Pax Christi Church parking lot. One woman, a media contact, agreed to stay behind at the Greenpeace office and phone the press at 9:00 o'clock, after receiving a signal from her colleague at the landfill that our blockade had begun. She also said she'd phone the Mayor before his morning coffee to let him know a group of Eden Prairie mothers and children were siting in a school bus blocking the Flying Cloud landfill entrance.

During the protest, the Greenpeace office would field all press calls while I stood at the site talking with reporters. Banners and signs would be brought in the bus by Greenpeace people. And Ben Gordon had already flown into Minneapolis, ready to participate in the hoopla. All systems were go for the President's Day Blockade.

That evening, I phoned the twelve mothers who volunteered to ride the bus, verifying their commitment. First call: Bonnie Swaim. She regretted to inform me she just didn't feel comfortable going through

121

with the protest.

"I'll pray for you tomorrow, Susan," she offered, as an alternate plan for herself.

Another women backed out due to her husband's objections. Half the original group withdrew. I tried to understand how these women felt, but my heart sank. They let me down and I cried. We were counting on them! How easy to stay home and let others take the risks! I wanted to run away from all this, too, but there was no place to go. We had decided our last best hope to turn the tide would be in this drastic action. After all, we tried nearly everything else.

The Russian poet Yevtushenko's words came to me now as I agonized over this decision. "Half measures can kill when on the brink of precipices, chafing in terror at the bit, we strain and foam because we cannot jump just halfway across."

Our blockade was a bold, tough measure of strength; it seemed right, but I still shook with fear at the thought of executing it. However, since we had decided to do it, it must be decisively carried out. Very tired and worried, I went to bed and prayed:

"God, quiet my mind and body so I may sleep. I badly need rest to cope with whatever tomorrow brings; anything can happen. And please, Lord, call out all the angels in heaven to watch over these mothers and children as together we take this difficult stand to protect our families and the fragile earth of which we're a part. Keep us safe from harm. Amen."

I awoke at least ten times during the night, but the day dawned clear, beautiful, and zero degrees. I jogged out to the mailbox in my usual fashion, in my robe and slippers, to grab the newspaper. I opened it outside and "Voila," Susan Varlamoff was staring at me on page one of the Metro Section, standing in front of the barbed wire fence at the dump. What timing! I ran back into the house with the newspaper. There was no time to read the article. This was B-Day.

Ivan got off to work while the boys and I ate a fast breakfast and dressed warmly for the day. The bus would be parked with its engine turned off, so there'd be no heat. The boys bundled up in their snowsuits, scarves, mittens, hats, and boots. I dressed in wool pants, high leather boots, a sweater and turtle neck, my long Sunday red wool coat, a scarf, and Pierre's ski gloves. I tried avoiding the radical, disheveled look; I wanted to appear like a normal mother, except today I'd be standing in front of garbage trucks in the bitter cold.

After we were ready, Pierre, Neil, Paul, and I hopped into our blue station wagon and drove up Homeward Hills Road to Pax Christi Church at 8:30 a.m. We were the first ones to arrive. Since the church was locked, I ran the car motor and turned on the heat to stay warm. Minutes passed like hours. Would we be the sole occupants of the school bus? Would the rest of the mothers lose their nerve and stay home?

Finally, to my great relief, other cars carrying moms and kids streamed into the parking lot. Mary Cooper came with her three little boys, Mary Kennedy drove up with Lane and Paul, Barb Bohn and Randa Hahn arrived with their boys, and Jerri Coller came alone, deciding her children were better off at home with Grandma. Alex Zubar took the day off to be with us. He was a lone man among the bus load of women and children, so we gratefully welcomed him.

As people got out of their cars, I walked over to them and put my arm around their shoulders. They must be scared, I thought, because I was trembling. Everyone dressed warmly, and Mary Kennedy brought thermoses of hot chocolate and cookies for the children.

At the prearranged time, about 50 activists from Greenpeace, Earth First, and Stop the Burner Coalition rolled up in a school bus and separate cars. The bus emptied and the mothers and children climbed on board to set out for the landfill gate, just a few miles from the church.

Dead silence gripped the bus. Even the children stopped chattering. The only sound was the bus engine. This was certainly the longest bus trip of my life. Fear overtook every cell of my shaking body as the driver cautiously took back roads to our destination. He wanted to avoid an early detection from crossing the double lane highway in front of the landfill gate. Better to approach unexpected from another direction.

My brain shifted into overdrive again questioning the sanity of this decision. Would it work? Was it necessary? Could the dump trucks ram through the blockade? Was I risking the lives of these mothers and children? Would the children be warm enough to sit for hours in the cold? Was I prepared to be a good spokesperson? Just ahead the BFI sign came into view. It was too late to back out now. We must seize the day!

The bus driver made a sharp right hand turn through the gate and quickly positioned the vehicle across the entrance. He shut off the motor, and I jumped off clutching the expired landfill permit in my gloved hands. A Greenpeace person bounded out after me and the bus door was locked behind us. Then, the wheels of the bus were chained. Someone flew Doug Larsen's American Flag out the window. As we waited, the remaining Greenpeace people arrived in cars with various banners and signs to drape on the bus, cover up the BFI sign, hang down from the nearby drive-in screen, and hold up along the highway. These young people brilliantly executed this demonstration. They were professionals at this job. We watched stupefied as this whole scene magically unfolded before our eyes in a blur of coordinated activity.

Ben Gordon and several other men climbed on top of the bus with signs saying BFI, Big Fat Income, BFI Big Fat Indictment. Greenpeace mountain climbers slowly ascended the drive-in theater screen, trailing two long banners that read "BFI OUT," "Don't dump on us." Others grabbed signs and stood along the highway waving them. Another group of young men ran through the snow on a gentle hill and wrote in giant letters, "BFI OUT." Then they jumped up and down and laughed, very pleased with the result. What a spectacle! It was hilarious!

However, I was too afraid to be amused. My body visibly quaked now from cold and fear. After five minutes of this madcap scene, I loudly proclaimed, "This protest is a failure. The press hasn't come, no trucks have pulled up to the gate and BFI hasn't even noticed that we blockaded their entrance."

"Don't worry, Susan, it will all work out, you'll see," said Norm. "Just be patient."

He was so reassuring, but I still impatiently paced back and forth in the snow waiting for action. Not for long though. Noticing some commotion at his front gate, the landfill manager came storming out to investigate what was happening. He was a big, burly man of 55 years who resembled a football lineman. He demanded to know what we were doing. I showed him the February 28, 1988, expiration date on the permit and told him since BFI and the MPCA didn't shut the dump, we did it for them.

After a short shouting match, the manager stomped off, no doubt to call his supervisors. Now dump trucks began lining up behind the school bus, unable to get through the entrance. Many of the drivers got out of their vehicles to check out the problem. They laughed when they saw a school bus full of children munching on cookies and hanging out the windows talking to everyone. Meanwhile, Bill Schmidt drove back and forth on Flying Cloud Highway honking his horn and shouting out the window as he passed the landfill gate. Ahh. Sweet revenge.

Soon TV cameras, and radio, TV, and newspaper reporters swarmed the site like bees returning to the hive for honey. There were so many, I had trouble speaking with them all. Most of the time, I kept vigil under the American flag, repeating Thomas Jefferson's words again and again, explaining to reporters we mothers were following his vision of democracy by setting things right, because the situation was so far wrong. Yes, we had information which indicated the toxic landfill posed risks to our families, and we also realized the Minnesota government had every intention of expanding it. Therefore, we symbolically shut it down ourselves.

When Dean Rebuffoni's car pulled up, I jumped inside to get warm and give an interview. TV cameramen photographed the bus and even spoke with the kids. Lane Kennedy, three years old, said she wasn't sure what the uproar was about but big brother Paul, age seven, talked about groundwater pollution. Pierre, meanwhile, called out to radio reporters telling them, he was Susan Varlamoff's son. Well indoctrinated at home, Pierre listed the full range of problems at the dump.

On the other side of town, Sidney Pauly drove along complacently in her car listening to music, when an urgent message came on the air: "We interrupt this program to bring you this special announcement. A group of mothers and children are blockading the entrance to the Flying Cloud Landfill to protest the expansion...."

Back at the dump, what could have turned into disaster, became fun. The kids stayed warm and well fed, thanks to Mary Kennedy. The older boys thought the blockade was "awesome," all the action, "cool." Garbage truck drivers laughed in amusement at the long line of trucks stalled for blocks behind a school bus occupied by mothers and children. We all cheered when Greenpeace climbers reached the top of the theater screen and unfurled their giant banners. And the Greenpeace sign bearers were having the time of their lives, running through the snow hooting and hollering. We completely forgot about the possibility of being arrested. Besides, how could the police arrest this merry group of people?

The only problem I didn't foresee was frozen feet. My boot soles, obviously too thin, forced me to stand on my boot sides to keep my feet off the cold ground. Otherwise, all went perfect.

However, what we didn't know, but later heard from a reliable source, was that local police, state troopers, and a SWAT team sat mobilized in their cars two blocks away in case trouble erupted requiring their intervention.

After three hours of demonstrating, the landfill Manager ordered us off his property. To show he meant business, he got out his biggest truck, revved up the engine, and aimed it directly at the bus. Suddenly, our jolly group turned anxious.

"Unlock the chains, and get us out of here," we yelled to Norm and Ben.

The locks froze; it took some time to thaw them. Ten minutes later, as we prepared to exit, a reporter from Associated Press drove up to get the story and take pictures. We stalled a few more minutes to accommodate him.

Finally, we drove off, chanting "Recycle, Recycle" as we headed back to the church. Mary Kennedy, Jerri Coller and myself discussed repeating this protest for a longer period. Jerri thought a two day bus blockade might be fun. She suggested we set up shifts of people for a lengthier demonstration. We decided to keep this in mind.

Back at the church parking lot, I invited the 40 numb Greenpeace activists back to the house for hot soup and cider, bread and fruit. Norm questioned the seriousness of my intentions. Was I brain damaged from the cold making such a preposterous offer? I assured him, I was prepared to give everyone something to eat. These young people shouldn't go away cold and hungry. My mother perpetually fed everyone that came within range of her, so I guess I inherited this family character trait, too. For all the help these Greenpeace people gave us, a small lunch was such a pittance in return.

In a caravan, we all drove down Homeward Hills Road and turned right into Jack Pine Trail. Parked cars lined half the street. Then a steady stream of interestingly attired human beings filed into the house, took off their boots, all 40 pair, and tossed them onto our too small throw rugs in the foyer, completely mixing them up. I wondered how anyone would ever identify their boots when they had to leave. Then the boys and I took turns holding out our arms to accept their coats and run them up to the bedroom to deposit on my bed.

Our modest sized house quickly became overrun with humanity in various stages of frostbite. Bodies sprawled on chairs, couches, and on the floor, in the living room, dining room, kitchen and family room trying to warm up. The family room floor had the most takers. The sunny spot in front of the sliding glass doors which faces south, drew shivering bodies to it lined up shoulder to shoulder, like birds on a telephone wire. I laughed when I saw them.

Pierre, Neil, Paul and I quickly began the task of feeding our frigid guests. Pierre and I ladled out hot soup and cider while the younger children walked around with baskets of bread and fruit to offer everyone. The Alpine climbers received the first ration of soup from the crock pot. I was sure they expended the most energy scaling the drive-in theater screen in the bitter cold. One of these climbers, a polite Englishman, expressed enormous appreciation for the hot food.

Everyone behaved wonderfully and we debated various subjects at length while they ate. It proved to be a wonderful experience for the children. The Greenpeace people impressed the boys and me with their obvious intelligence and sincere concern for the global environment.

During the food distribution, Ben Gordon stationed himself at the dining room table with the telephone. He called his office to update the press person on events and give our phone number. For the remainder of the afternoon, our phone rang nonstop with reporters wanting to interview me. Thank goodness we had call waiting. When phone conversations were interrupted by other calls, I'd ask the person on the second line to get back to me in ten minutes, the next caller in twenty minutes, and so on.

Grant Merritt saw us on the news and somehow managed to get through on my line to inquire how we survived the morning. "Fine," I said, but physically I grew weary. By about 3:00 p.m., I laid down with my back flat on the floor and my ear glued to the cordless phone speaking to the press. The doorbell rang constantly, too. One man tried selling me pollution control devices, and an Indian proposed to end all our worries by taking the county garbage on his reservation. Even Ivan's company phone line jammed with calls, because they mentioned it in my feature story and people called him to track me down.

By eight o'clock, I crashed in front of our TV and fell sound asleep in Ivan's arms. I pitied my poor husband having to deal with a rabble-rousing wife. Barely had I fallen asleep when the doorbell rang. Grouchy, I reluctantly awoke to answer the door. Standing there was our wonderful Greenpeace bus driver smiling and asking for a donation. I was furious. In my mind, my generous offer to feed these people should have been enough, but now these people wanted money, too. I showed visible annoyance and gave only a small donation. Later, I regretted that stupid decision and realized fatigue caused my poor judgement. How unfair of me to be so cheap! These young people worked hard and well for us, and needed the money to pay their rent. It still upsets me to think of what I did.

We achieved our goal. The story of the Eden Prairie mothers and children blockading the front entrance of the Flying Cloud Landfill made headline news on all Minnesota television and radio stations, local and state newspapers, and even *U.S.A. Today.* Now that we had everyone's attention, we needed to run like gazelles around government offices and demand action on the issue. Maybe Minnesota officials would take us seriously this time.

7 | SHUT DOWN

A near capacity crowd of angry citizens packed a meeting room at the Eden Prairie shopping mall to demand a contested case hearing on February 17, 1988. Representatives of MPCA's Board, MPCA's staff, and BFI's finest, faced off against representatives of the Eden Prairie City Council, city attorneys, environmental lawyer Grant Merritt, Representative Sidney Pauly, Leslie Davis, Greenpeace activists, and various concerned citizens. We brought our best guns to open fire on the opposition.

As I recall, we first arranged to have neighborhood children explain the recycling banner they designed to cover the BFI entry sign the day of our blockade. Janet Green, MPCA Board member and Chair for the evening meeting, expressed her admiration for the children's artwork. "This is the best recycling banner I've ever seen," she said. When our young narrator drew attention to the sun sticking its tongue out at the BFI facility, the room dissolved in laughter.

After the children finished, I took the microphone, pointed my loaded tongue at the MPCA and blasted away. I demanded to know why

129

MPCA's staff displayed outright negligence in dealing with BFI's many violations. We repeatedly appealed to the staff for help, I pointed out, and recently asked them to shut down the dump when its permit expired. Our pleas fell on deaf ears.

"We condemn this landfill expansion," I said. "Ninety-nine percent of the people in this room know the landfill will come back to haunt us."

As I talked, I became angrier and angrier. I watched my clenched fist go up in the air and slam down on the podium again and again and again. Forces beyond my control seized my psyche. I felt like an innocent bystander to a spectacle that someone else inside me created. Perhaps my unconscious mind urged repeated banging on the podium to hammer my words into the minds of government officials. I held nothing back. Never before did I get so upset publicly, but I knew we had to convince MPCA's board to grant us a contested case hearing, and this was the opportunity to do it.

Odds against us were enormous. The Metropolitan Council had already voted overwhelmingly to recommend BFI's landfill expansion, and MPCA staff advocated the same. The permit only required MPCA Board approval before bulldozers could start digging.

I had nothing to lose by acting enraged. It was how I felt anyway, angry at a process that victimized us for the almighty buck and the comfort of having an operating landfill within the Twin Cities Metropolitan area.

Many other citizens stood up and verbally charged representatives of the MPCA and BFI for lack of integrity. Greenpeace activists stationed themselves around the room jeering and heckling to excite the crowd. The room heated with this angry mass of agitated humanity.

Ben Gordon took a turn at the microphone to tell the MPCA, "The contempt with which BFI has trampled over the people in this room astounds me." Since the company's criminal record increased significantly since we last testified, he announced one by one, BFI's current crimes. Ben discussed a grand jury investigation of BFI by the U.S. Department of Justice, and convictions for price-fixing in Ohio, New Jersey, and Georgia.

When our favorite BFI lawyer got up to speak, word traveled down the rows of people to walk out of the room. Our diminutive State Representative Sidney Pauly, led the troops in parade formation out the door. Poor Dick spoke to empty chairs. Later, we learned from the newspaper that Dick noted the company hadn't been charged with any wrongdoing in Minnesota, and "BFI is not an evil corporation." Good thing we left, because we would have caused an uproar at these remarks.

Ric Rosow spoke next from his seat in the audience.

"Two weeks ago I wrote the MPCA, and am chagrined the MPCA has yet to answer my letter. I read the MPCA's response the other day in the Minneapolis *Star Tribune.*"

Listening to all this brouhaha but saying nothing was Councilman Dick Anderson, a "Ring Around the Dump" veteran. He finally walked up to the microphone and stunned the room.

"I was sickened to find out that neighborhood mothers and children blockaded the landfill in an act of desperation. This clearly showed they lacked faith in the governmental process to protect them."

I quickly stole the moment to run back to the microphone and announce the terror I experienced worrying about the safety of my friends and their children that day.

"We risked our necks because you wouldn't help us," I shouted. Then I looked at the panel of Board members and demanded, "Would any of you be willing to live in our homes by the landfill?"

Dead silence. No volunteers.

Visibly mad, Roger then defiantly marched up to the podium and proclaimed: "This community is concerned that no one listens to its complaints, and there's a lack of confidence in the review. This matter should go to a contested case hearing."

We shouted our approval.

This was brilliant strategy. Roger waited for the citizens to get upset, then jumped on the moment to make his demands known. Timing par excellence.

Janet Green finished the hearing by saying, "The first decision we will make will be on the contested case hearing."

The meeting broke up after two-and-a-half hours and we all milled around talking to MPCA officials to get our message across one last time. Cary Cooper spoke at length with Janet Green telling her, "I'm sick and tired of BFI falling through the cracks in the regulations."

Janet said she heard Cary loud and clear and promised to get a handle on this situation. That night, MPCA Board members must have felt like fish swimming in a piranha tank, ready to be eaten alive.

After another long, exhausting evening, I drove home and entered a quiet house. The family slept soundly as I dragged in late once again, took off my coat, and trudged up the stairs to bed. How many more nights would I have to spend like this?

The Attorney General takes a stand

We desperately needed to corral more political support to force the hand of the MPCA board to grant the contested case hearing. Grant came up with an idea. Hubert H. Humphrey lll was his next door neighbor. So what, you may ask? Well, this former Vice-President's son also happened to be the Attorney General of Minnesota. Friends called him Skip. We called him Skippy, like the peanut butter. We had tried in vain to contact his office many times, first by letter, then by phone. There was no answer. Our last effort was made the day of the blockade, when we mailed a letter to Skip's office with a copy to the press. Still, no answer.

Grant finally telephoned his neighbor at home to advise him to contact Susan Varlamoff, spokesperson of the Flying Cloud landfill anti-expansion group. Grant told Skip this landfill issue is a political powder keg. He must address the concerns of people or face political consequences.

One evening about nine o'clock, the phone rang. Ivan answered it and handed it over to me. I said hello and the voice on the other end said, "Hi, this is Skip Humphrey. Grant tells me you need some help with this landfill problem."

Skip spoke to me like the guy next door wanting to know how he could help his neighbor in distress. I lost no time beginning my soliloquy of frustrations dealing with the Metropolitan Council and the MPCA on this issue. Then I asked if I might speak with him in person to discuss the case.

"Call my office in the morning and make an appointment,"Skip replied. A political door had opened.

Wasting no time, I called Skip's office the next day after the children left for school, and set a date to meet with him. However, before I traveled to the state capitol carting my piles of papers, I spent considerable time deciding what to wear for this momentous occasion. Grant told me Skip was short, so high heels were out. I didn't want to overpower him. I tried portraying the image of a vulnerable mother fighting for the welfare of her children, which in fact I was.

Grant, on the other hand, played the more aggressive role with his loud, gruff voice getting the attention of the person or group we were addressing. He prepared the way for me to speak. This complementary relationship was natural. Grant felt confident because he knew many government people, having once been in the arena himself, but I was totally intimidated, probably reflecting fear in my manner. This relationship worked like a charm, so we stuck with it.

For my attire, I chose a long, black, pleated skirt, white blouse, and long, colorful, fringed shawl I bought that past summer in Belgium. I thought the shawl might give my face a little color. My skin was ghostly pale in winter, and this year, it was even whiter from the fatigue of leading our quest.

Grant and I agreed to meet in the waiting room of Skip's office. After we arrived, we gave our names to the secretary and waited to be called to present our case. I sat uncomfortably at the edge of my seat while Grant, as usual, found someone to talk with. Eventually, we were ushered into the Attorney General's office. The decor of the room was superb. Fully paneled walls held beautiful objects of arts. I walked along the walls inspecting them until an assistant to the Attorney General joined us. She carefully listened to my story, backed up by reams of documents, and suggested perhaps I might want to become a lawyer. (I'd give it some thought if encyclopedia saleswoman didn't work out.) She agreed there were serious concerns with the landfill expansion.

Then Skip entered the room accompanied by his assistant and I stayed seated. I had the impression I'd tower over him even in flat shoes if I stood up. After Grant introduced me and we shook hands, Skip remained standing several moments, merrily expounding on some current crises. I found him very likable.

I restated the case in short, and Grant held open a document with phrases underlined regarding some court case involving BFI. Skip nodded in agreement. I added that neighborhood mothers, in an act of desperation, blockaded the toxic landfill because we received no help from the government to protect our families against its expansion. "I hope we never have to do that again, it's embarrassing," I added.

Skip let out an uncomfortable laugh.

Grant discussed the lack of concern, or perhaps collusion, between BFI and governmental agencies to push through the expansion. "Giving this company a landfill monopoly would be big trouble for Minnesota," he warned.

Grant discussed the MPCA's dismal record of failing to force BFI to comply with regulations, and suggested a grand jury investigation of the matter.

After very little reflection, the Attorney General consented to recommend a contested case hearing to the MPCA Board. "A copy of the letter will come in the mail," he said. Grant appeared tongue-tied, a rare occurrence. He stuttered a bit with this fast pace of events, and profusely thanked his neighbor. I warmly shook his hand in appreciation. Grant and I left the office flabbergasted at our good fortune.

The bandwagon fills up

Several days later, I drove out to the Minnesota branch headquarters of the U.S. Fish and Wildlife Service, and point-blank asked Ed Crozier, the manager, for a letter recommending a contested case hearing. With no hesitation, he agreed to write one, based on unanswered technical issues regarding the impact of the landfill expansion on a wildlife refuge.

When both letters arrived, I got back in my car, headed downtown to the Minneapolis *Star Tribune*, and left them for Dean Rebuffoni. Next day, the newspaper featured an article with Skip's picture, announcing the Attorney General and the U.S. Fish and Wildlife Service requested a contested case hearing in the Flying Cloud Landfill case. What a coup! Could the MPCA board refuse to grant the wishes of the Attorney General? Never!

Next on my lobby list was Mark Andrews, head of the Hennepin County Board. Mark had sponsored a resolution the past December to temporarily expand BFI's Flying Cloud Landfill while awaiting approval of a permit. BFI officials convinced him this was the best option for securing additional landfill space needed in the Twin Cities. I thought if Mark heard the other half of the story he might see the situation differently.

Before I traveled downtown to speak with Mark, I researched the profile of this young, ambitious politician featured frequently in the Minneapolis *Star Tribune* for his exploits. In pictures, he appeared tall, thin, and good looking. He was my age, a married man, and the father of two young children. He attended college in Minnesota, and ran his own french fry stand at the state fair every summer. Mark and Grant worked together on environmental issues many years earlier, and Grant spoke highly of him. Currently, an uproar raged over the Minneapolis incinerator proposed by Mark as an alternative to the landfill disposal of garbage. This caused him a lot of grief, as did the landfill expansion cause for me. Maybe we could find common ground to address the Twin Cities' solid waste problem by helping each other.

As I waited outside Mark's office, he finished a phone call, giving me time to collect myself and my thoughts. When he came out to greet me, he apologized for his tardiness and invited me into his office. Once inside, I admired the magnificent city skyline from the impressive surroundings of his large office. The decor was pale blue with comfortable sofas placed in groupings for a relaxed atmosphere. A massive wooden desk stood at an angle in one corner with chairs opposite. Turning to Mark, I shook hands and expressed my gratitude for the opportunity to look into the face of a famous political figure. "I read about you so often in the newspapers," I said.

Mark laughed easily and said he was pleased to meet me, too, having seen me frequently on television. We were on equal ground, I guess, except he earned big money for his hard work, and I received nothing for mine.

Mark looked better than his Minneapolis *Star Tribune* photos. His smile radiated warmth, and his thin frame resembled a young boy's lankiness. However, his appearance didn't fool me. Mark was a powerful person in Minnesota, and this was the reason I had come calling.

"Beware of the lean and hungry look," cautioned Shakespeare in his play, *Julius Caesar*. Mark had this demeanor, the look of a very ambitious man. Better this energy be with us than against us. By showing him the underlying environmental problems of BFI's landfill expansion, I hoped to persuade him to change his mind on the issue and support us.

Mark offered me a chair directly across from him, which I accepted. I then proceeded with my discourse, one I could now recite in my sleep. Mark's eyes sharply focused on me as he listened attentively to my every word. Rarely have I felt such intensity. He never uttered one word and never lost his concentration for one instant as I spoke. To back up my opinions, I provided xeroxed copies of pertinent information. He quickly scanned them, appearing to absorb everything while I silently watched.

Then Mark spoke. He conceded he had heard only BFI's version of the landfill expansion story from the company's representatives that swarmed the town, lobbying for their "hi-tech" expansion. Mark agreed we had legitimate concerns. With this new information, he believed the county's solid waste program might require revision.

Landfilling ranks rock-bottom on the priority list of solid waste management plans, Mark said. He expressed his desire to eliminate the need for massive state landfills with the big incinerator he proposed for downtown. He showed me specifications and designs of bag house filters and electrostatic precipitators. The vocabulary shot over my head, but I strained to appear knowledgeable and interested.

Mark obviously wanted my endorsement for his incinerator, and I needed his condemnation of the dump expansion. I assured him I wouldn't oppose the incinerator, because my landfill protests already occupied all my time and energy. The arrangement worked well. Mark no longer advocated the Flying Cloud Landfill expansion, and even went so far as to lobby against it in the legislature. In turn, I steered away from opposing the incinerator. We worked in cooperation after this meeting, an arrangement, I believe, we mutually enjoyed.

Head on collision at the MPCA

Rod Massey, MPCA Solid Waste Division head, came after Mark Andrews on my contact list. I called his office, got an appointment, reorganized my papers, and pointed my car toward St. Paul and the MPCA. Rod Massey escorted me into a conference room containing a long table which provided me ample space to spread out my library of dump documents. I noted how the drab, yellow-painted walls, chipped in places, contrasted with the posh, paneled surroundings of the politicians' offices. It was shocking and depressing. Evidently, our tax money wasn't spent on pollution control offices.

My presentation to Rod started off slowly and pleasantly, but gradually accelerated in pace, and darkened in mood. I cited violation after violation that occurred at the Flying Cloud landfill, followed by each of the MPCA's responses.

BFI broke the terms of the original permit by operating the facility

136

24 hours a day, seven days a week, and by accepting hazardous materials; the MPCA rarely monitored their activities. BFI doubled its landfill capacity without permission; the MPCA wrote an amended permit to include the expansion and add 10% additional capacity. BFI withheld monitoring well data showing groundwater contamination for two months; no violation was issued. BFI promised to shut the operation in 1982 and asked for an expansion instead; the MPCA recommended an expansion. Fires and an explosion occurred at the site; no violation was issued. BFI failed to meet the deadline for the landfill final cover; the MPCA granted an extension. Now that the landfill permit expired, the MPCA searched for reasons to allow BFI to continue operating. Every point I made, I supported with documents. I breathlessly finished my long list at roller coaster speed, reminding the solid waste manager that most likely these violations represented the tip of an iceberg. So much information was lost when the dump files were tossed out in 1980. We had no idea how much or what kinds of waste lay buried in the pit, I said.

"Why did you allow this facility to go unmonitored for so many years?" I demanded to know. "And how could you support the expansion of a toxic landfill?"

In contrast to Mark's intense, deep gaze into my eyes, Rod fixed his eyes on the conference table as I spoke. That struck me. Why couldn't he look at me? As I continued on about BFI's negative corporate profile, Rod cut me off. He told me he thought BFI officials were easier to deal with than many others. They were always ready to remedy a problem, once it was brought to their attention.

What a clever gimmick; another of my children's ploys, I thought. After my sons liberated the cookie jar of its contents, without permission, they'd always show great remorse. Did I allow them to go unpunished? Never. What's a small apology next to a belly full of scrumptious cookies or fat corporate profit. No wonder Dick could confidently strut around town saying BFI had few violations in the state. He was right. When it broke the law, the MPCA let it off the hook or abetted its operation by adjusting the permit.

My conversation with Rod Massey headed straight into an abyss. I warned him we wouldn't tolerate BFI's expansion and wanted its landfill shut down.

On my way out of the building, I made a little detour to the top floor.

137

I walked up to Commissioner Willet's secretary and politely asked for a date and time to meet with her boss. A little taken back, she accommodated my request. I left the MPCA pleased to have secured a meeting with Willet, and not at all surprised I was unable to change Rod's thinking on BFI. Hopefully, I impressed upon him our strong desire to see the permit expiration date enforced. The bus blockade should have made him realize we meant business.

Lobbying in and out of town

I dropped by Grant's office often to brief him on the latest events. We wanted to keep our lobbying quiet and confidential. To avoid paying parking fees downtown, I parked in the Catholic church lot. A sign on the fence said the area was reserved for church members only. Feeling guilty, a Catholic quality, I always entered the church for a five-minute prayer to justify my use of the space before going on to Grant's office. In my purse, I carried prayers which I recited. During this period, when a powerful Minnesota government agency and BFI pummeled us mercilessly, my thoughts often turned to these words of Thomas Merton:

> *"My Lord God, I have no idea where I am going.*
> *I do not see the road ahead of me. I cannot know*
> *where it will end . . . And I know if I will do*
> *this you will lead me by the right road, though I*
> *may know nothing about it. Therefore, I will*
> *trust you always, though I may seem to be*
> *lost in the shadow of death. I will not fear, for*
> *you are ever with me, and you will never leave*
> *me to face my perils alone."* Amen.

Feeling emotionally rejuvenated, I continued on to Grant's office. Our lawyers were now working to draft a request for a contested case hearing. MPCA regulations required us to identify specific issues of fact for the hearing. Grant, Ric, and Roger explored this at length. Eden Prairie citizens busily distributed information flyers and wrote letters to the editors of all the newspapers keeping up the heat. Even my brother John actively supported our cause in Denver, Colorado.

One evening, while John entertained clients in a Denver hotel bar, he noticed a man wearing a BFI cap. Unable to resist an opportunity to poke a little fun, John strolled over to this man and his companion and sat down beside them to question them about the solid waste business. They emphatically declared the garbage industry was booming; BFI's stock soared. My brother then casually mentioned he had heard the company was experiencing problems at some of their landfills. Yes, this was true, they said. In fact, they were flying to Eden Prairie, Minnesota, in the morning to do a little trouble shooting.

My brother then remarked: "You might know my sister, she lives in Eden Prairie."

Wide-eyed, they asked, "Who is your sister?"

"Susan Varlamoff"

Red-faced and angry, the two men got up to leave. One said, "you tell your sister that BFI will get the f- - - - - - expansion no matter what."

As they stormed off, my brother yelled after them, "Not if my sister is involved."

My brother couldn't reach me, so he called my mother to tell her this incredible story. My mother then phoned me to relay the hand-me-down news. I then telephoned my brother to hear the story first hand. We laughed and laughed. Life was getting so hectic, I appreciated this little reprieve.

● ● ●

Reading the newspaper each day was an act of courage. One never knew what bad tidings lurked inside, so when I opened it, I steeled my nerves for the worst. In the spring of 1988, the headlines, PANEL VOTES TO END SEARCH FOR LANDFILLS, set us reeling again. Dean Rebuffoni reported that a Minnesota subcommittee approved a bill to drop eight large tracts for consideration as potential landfill sites.

I reasoned if the landfill siting process was stopped, then Flying Cloud's expansion would be inevitable. Our war opened on another front: the legislature.

I called Sidney Pauly and Don Storm and cried out for help. They hadn't heard about this bill. Thank goodness for Dean's article. Both assured me they would get right to work and look into the matter. Then I spoke to Grant and he said we had to hustle over to the capitol and lobby against the bill. God in heaven, what did I know about lobbying in the legislature. Something else new and scary to do. I was already so tired. I didn't know how I would cope with yet another crisis.

Sidney quickly prepared packets of information containing BFI's corporate record, a chronology of the dump, and scientific reports from Linda which she passed out to her colleagues. This bill, which originated in the Senate, first had to pass through several subcommittees before the full Senate voted on it. Then the bill would travel through the House plumbing for its first hearing before the Environment Committee. Sidney sat on this committee.

In another bill, we discovered a clause that would prohibit the city from using their zoning powers to block the landfill expansion. It, too, originated in the Senate. Sidney telephoned Don Storm to explain the situation. Don lost no time finding his colleague who sponsored the clause, and gave him a verbal thrashing in a back room. The clause was withdrawn. After this, we learned to be vigilant for any eleventh hour trickery.

Gene Merriam, nicknamed "Clean Gene" by his colleagues for his impeccable record, sponsored the amendment to abolish the landfill siting process. The newspapers quoted him saying "If everything goes right we aren't going to need any more new landfill capacity."

Could his "everything goes right" remark mean he expected BFI's Flying Cloud Landfill expansion to be approved? Did he realize the problems with this expansion? I called Gene's secretary to make an appointment to educate the Senator on the hazards of enlarging BFI's Flying Cloud landfill.

In the meantime, Grant and I arranged to meet at the Minnesota State Capitol to lobby subcommittee members on the Merriam Amendment. At the designated place and time, I awaited Grant. He was running late. Tired, I leaned against a concrete column to daydream as I watched the activity around me.

Men and women scampered around the capitol building's columns like squirrels in a forest. An anxious din of chatter filled the air. Everyone searched for or gathered information like nuts or seeds they might eventually plant to bear fruit for some company or special interest group.

The laws of Minnesota seemed irrelevant as these well-dressed men and women plied their trade. Signals passed across rooms between politicians and these influence peddlers as bills were discussed. Meanwhile, reporters dashed around eagerly, pads and TV cameras accompanying them, hoping to scoop a story. I felt I had entered some exotic land for all I understood about the culture surrounding me. It made me wonder what our country's founding fathers would think if they awoke into modern times like latter day Rip Van Winkles. How would they feel about representatives of special interest groups choking the halls of our once sacred legislature, influencing votes?

Grant finally arrived, invading my fantasyland of squirrel images. We were both hungry, so he bought us apples at a little fruit stand. We munched on them while we hurried over to a subcommittee hearing on our bill. Public testimony wasn't permitted, but Grant and I positioned ourselves in a prominent place to show we were observing the proceedings. At one point, Grant left his seat to lobby politicians outside the room, appointing me guardian of his briefcase. I warned him, "Don't be gone long. This place makes me nervous."

The proceedings and people were hurried, and all this new stimulation overwhelmed me. Thank goodness we couldn't testify. I could never have stood before this group and made any sense. While Grant was gone, Dean Rebuffoni walked in and sat next to me. (He probably saw the terror in my eyes and pitied me.) I expressed to him my anxiety over Grant's long disappearance. Dean laughed.

"Grant is out lobbying," he said. "That's what he's paid for, isn't it?" Then he laughed again.

I wasn't the least bit amused with his comment, but I attempted to look cool and calm, clutching my sweaty palms together while my heart beat frantically. I was about as relaxed as a wound spring.

We lost the vote in this subcommittee. However, Grant assured me time remained to kill the bill, since it had a long route to follow through other committees before the full Senate voted on it. Until now, BFI had always convinced politicians to push through its agenda in the legislature, but we were determined to halt the pro-corporate mentality.

We attempted to speak before several other subcommittees and failed. Every time the chair recognized my face at a meeting, the agenda was rearranged, and the bill containing the Merriam Amendment postponed for a later time or another day. But when I was absent, it would be passed. I sensed a conspiracy.

In every Senate subcommittee, we lost the vote. Finally, the bill went before the full Senate. The meeting was closed door. No one was permitted to testify but senators. Loudspeakers above Senate doors gave interested people standing outside the room the opportunity to listen to the proceedings. Randy Johnson and I stood helplessly outside the great wooden doors listening to Don Storm heroically appeal to his colleagues to vote down the Merriam Amendment. In a small effort to aid him, Randy and I wrote down dump data on scraps of paper that senate pages brought to Don while he spoke.

The room exploded in a fury of angry discussion after Don finished his remarks. No senators wanted a landfill in their district and frankly, they didn't care what problems existed at the Flying Cloud Landfill. Despite Don's heroic effort to change the senators' minds, they voted overwhelmingly against us.

During these days of tumult, I tried speaking with Senator Merriam. He became occupied with other matters and canceled his appointment with me. Gerald Willet had a conflict with our meeting, too, so he bowed out. Certainly, with the legislature in session, I could understand how busy these men must be.

Undaunted, I called both of them for second appointments. Linda Lehman also telephoned Merriam's secretary for a meeting. These were canceled too. Now my ire was up. Obviously these gentlemen were avoiding us. I re-phoned Willet's secretary and carefully explained I needed to speak with the commissioner and must have an appointment. She gave me one. I quietly told her to relay this message to the MPCA director, "Tell Mr. Willet he must keep this meeting with me. If he doesn't, so help me God, I'll find him at the appointed time, even if I have to go in the bathroom after him." She agreed to tell him.

March 18, 1988, Grant, Linda, Ric and I confidently marched into a nondescript, windowless conference room at the MPCA and sat down on one side of a long table. On the opposite side sat MPCA staff members including their lawyer. As I recall, there were four other people besides Willet in the room. Commissioner Willet sat at the head of the table.

I introduced myself to the commissioner and announced that I brought along a few friends for this momentous occasion: two lawyers and a scientist. We all shook hands and sat down. I took the chair beside the commissioner and pulled my seat as close to him as I dared, without falling into his lap. I dressed in the same black skirt, white blouse, and colored shawl I wore for the Attorney General. Maybe it would bring me good luck again.

The room quieted and I began my presentation. I turned my body to face the commissioner and looked directly into his eyes without flinching. They looked empty, nearly dead. There was no light in them. Staring into this expressionless face, I quietly and systematically set out the facts in the case and again expressed distress the MPCA had been grossly negligent in dealing with BFI.

"We are sick and tired of this treatment," I said.

After my ten-minute monologue, Linda spoke. She turned page after page in a large notebook containing methane data, and firmly pointed out that the gas was leaking off the landfill site into the Homeward Hills neighborhood at above explosive levels. Off-site methane migration violated MPCA regulations, because it was a safety hazard. (Earlier in the year, the MPCA asked Browning Ferris Industries to install an active methane extraction system, but BFI neglected to act on this request.)

Grant now took a turn to speak. He got out of his chair and paced back and forth in the room like a caged lion. He was roaring mad and growled at Willet, "You don't have the guts to prosecute the big polluters. You just pick on the little guys," he said. Grant continued in this manner for ten minutes, and my ears just curled with his insulting remarks. Everyone sat silently listening to Grant's voice booming throughout this small room.

Having once occupied Willet's seat, Grant recognized the drawbacks involved in prosecuting big business. However, he felt that they did more damage to the environment than the corner gas station with its small oil spill, so he aggressively went after them during his tenure at the MPCA. He paid dearly for his behavior. His law firm would not accept him back after his government days because several lawyers insisted his presence would discourage big business from enlisting the firm's services. But everyone I spoke with ranked Grant Merritt as one of the best MPCA directors.

Ric made no comments after Grant's performance, but then why bother, nothing could top his act. Next, we discussed the methane gas migration. The MPCA lawyer said it would be dealt with fairly. I didn't believe a word of what she said and told her so. "We have yet to see the MPCA prosecute BFI for anything." Willet sat stone-faced barely uttering a word during the entire meeting. That worried me. Surely, he must feel something with all this agitated humanity in the room.

My recollections of this afternoon remain foggy. I was dead tired. In the weeks following the bus blockade, I pushed myself as hard as humanly possible to speak with as many high government officials as I could. My body was running on empty now. When I become very fatigued, my blood pressure drops, I get dizzy and feel like a limp rag. I had reached this point.

After approximately one hour, Willet adjourned the seance. As we prepared to leave, I panicked. This was our last opportunity to influence the MPCA head, so in my closing remarks, I grabbed Willet's arm just inches from me and told him, "We want to see a decision made in favor of the people. Shut the dump. We've had it. If you don't shut the landfill, I promise you the school bus will reappear in front of the Flying Cloud Landfill entrance."

As Willet rose to leave, he put his hand on my shoulder and said, "Don't get so excited about this whole thing."

"I have good reasons to get excited," I replied, "my family and friends' families are in danger, and the MPCA is doing nothing to help us."

Our small entourage left the room and stopped by the agency cafeteria for something to drink. While Grant and Linda bought drinks, Ric and I held a table. I told Ric we had wasted our time. Willet showed no reaction to our presentation.

"How could you be so sure he won't do anything?" Ric questioned. The truth was, I was simply too tired to think and see clearly. Very discouraged, I drank something, I don't remember what, and drove home.

How I made it home in my exhausted state remains a mystery. I could hardly see the road ahead of me. My good friend Brenda Gowan was waiting for me with her children, Jeff and Cindy, ready to eat pizza. Often, when Ivan traveled out of town, Brenda, a single Mom, and I shared dinner on Friday nights. The menu was always the same: pizza and cherry coke. There was no cooking, and the cleanup consisted of throwing out dirty paper plates and boxes, giving us ladies more time for our favorite pastime, talking.

After we fed the children, I sank into the sofa to listen to the latest outrage in Brenda's ugly divorce. My body begged for sleep. I tried to force my eyes open and listen to my friend's conversation, but nature won out. Several times my head nodded during the course of our conservation.

That weekend I was very depressed. It seemed idiotic to continue this crazy battle against BFI and the Minnesota Government. We were getting nowhere, I thought. I called Grant in near tears. I confided in him my complete frustration with the situation and expressed my hesitation to continue the fight.

Grant said, "You knew this would be an uphill fight, Susan. Be at the MPCA Board meeting next Monday. I have a hunch something might happen." Although still upset, I promised to meet Grant at 9:00 a.m. sharp, downtown at the MPCA.

Monday morning dawned a clear, crisp day. I got the boys off to school, put on my red Easter suit and drove our blue station wagon jeep to the Minnesota Pollution Control Agency. In the basement board room, Grant was excitedly talking to MPCA officials. There was an unusually high number of BFI lawyers attending the meeting. Ric arrived and the three of us sat down in front of the Board at a long, narrow, table with microphones. Ric sat on one side of me and Grant on the other. I felt well protected.

Grant scribbled me a note on his legal pad. It read: "We're on track here. I talked to Keith Langmo and Gerald Willet. What they are trying to do is 100%." I gasped in disbelief.

The chairperson called the meeting to order and Rod Massey stood up at the podium and spoke first. He read the following from his memo. "The Minnesota Pollution Control Agency finds Browning Ferris in violation of their permit No. SW-14 regarding methane control, and therefore orders them to cease operations at the landfill as of April 4, 1988."

Stunned, my body charged with adrenaline trying to grasp the full impact of these few words. I thought I'd explode in my seat. Grant coolly grabbed the microphone first and thanked the Board for doing even more than we had requested. I followed his lead and profusely thanked the MPCA Board also. After a few more minutes of discussion, we prepared

145

to leave the room. Trying to keep my composure took all my energy. I honestly wanted to stick out my tongue at those BFI guys and say, "Na Na boo boo, stick your head in doo doo!" but, instead, I acted maturely, concealed my glee, and stifled my immature thoughts.

Calmly, but smiling, Grant, Ric, and I got up and slowly walked to the back of the room toward the double doors where a grim, speechless, group of BFI men stood. Grant and Rick flanked me as we approached them. They parted like the Red Sea to let us pass and we felt like Moses leading the people to the promised land. Quietly, we turned to shut the door behind us. Then I grabbed both these men's shoulders and screamed, "We did it, we did it."

Ric mentioned that his letter to the MPCA probably forced them to shut the landfill. Grant and I agreed it was probably that.

Chuck Laszewski, a reporter for the St. Paul Pioneer Press newspaper, trailed after us to get my reaction to the MPCA's order. I replied, "We've been fighting this expansion for six years. We're finally seeing the top of the mountain and it's glorious."

Briefly we enjoyed the moment, then Grant hurried me over to the legislature to testify before the House Environment Committee. The session was just beginning, and we wanted to tell Sidney Pauly the news before she left her office to sit on the committee. Like two children running to meet their best friend on the playground, Grant and I raced down the halls to find her. We just made it. When I reached Sidney, I grabbed her tiny shoulders and squeezed them while I screamed the news into her face. She seemed beside herself with my outburst, which was not surprising. I nearly broke her shoulders and eardrums within 30 seconds of my arrival. Nevertheless, she recovered and was thrilled with our good tidings. Together we confidently strode over to the legislature ready to defeat the bill.

While Sidney took her seat on the committee, Grant and I looked for seats in the peanut gallery. Before we sat down, we searched the audience for friend or foe with which to share the breaking news. Only foes were in sight, so we approached them.

Grant saw Jim Erickson, a BFI lobbyist, and asked him what he was doing there. The room crawled with lawyers from Browning Ferris Industries and Waste Management, Inc. His face turned beet red as he explained he was just observing the proceedings. "Oh sure you are," replied Grant sarcastically.

I spied Josephine Nunn, Chair of the Metropolitan Council Environmental Resources Committee. Ecstatic, I blurted out the MPCA decision. Her face dropped. She was shocked and said, "This can't be a final shutdown, just a temporary one." I explained the shutdown was final until the MPCA granted BFI a permit for the expansion. Josephine Nunn was beside herself with this turn of events.

Grant returned with cups of tea and we took seats behind Miss Nunn to watch the meeting. He was in a bullish mood after the MPCA decision and told the chairwoman what he thought of her agency. "It should be abolished," he said in a gruff voice. A fire lit in Josephine Nunn's eyes. Her cup ranneth over. She would not forget this insult. She would get even.

Public comments were permitted at this hearing, so Grant requested we be given the opportunity to speak. Each speaker was allotted ten minutes, except us. We were given one minute each after everyone else testified. Somehow I got the feeling we were *personae non gratae*. We started arguing about this unfairness, but the meeting began. There was no time to change anything. Grant and I secretly decided to take the same time as everyone else.

The Merriam Amendment topped the agenda. A Committee member mentioned this bill would have a direct impact on the Flying Cloud Landfill in Sidney Pauly's district. A few guffaws and snickers followed. Sidney, dressed in a pale pink suit, slowly leaned back in her seat, rested her head on the plush chair back, and sweetly announced, "I was just told that the MPCA ordered the Flying Cloud Landfill shut down this morning." Silence. The only ones left smirking were Grant, Sidney and me.

The testimony began. Person after person discussed how a landfill would adversely impact their communities if the siting process continued. (Several towns were targeted as potential sites.) Then came our turn. Grant began first. He told the House Environment and Natural Resources Committee that this bill would put additional pressure on officials to expand BFI's Flying Cloud Landfill. He said the bill would "play into the hands" of Browning Ferris and Waste Management, Inc. Both companies he called "co-conspirators" noting they were accused of price-fixing and bid-rigging in Ohio, and paid a $1 million fine there.

147

Grant said lobbyists from both companies were in the room right now. "I'm not saying Representative Darby Nelson introduced this bill because they asked him to," he said. "He didn't, as far as I know, but. . ." Jeers and whistles interrupted his speech. Rep. Jean Wagenius rapped the gavel. "You're out of order," she said to Grant. A shouting match ensued. Several representatives felt Grant had insinuated they were influenced by these two big garbage companies. One legislator shouted at Grant to "Shut up." The chairwoman told him his testimony was over. Grant obeyed and angrily charged back to his seat beside me.

I heard my name called next. I was the last person to speak. This was my debut before the legislature. After Grant's attention-grabbing performance, all eyes and ears focused on me. I determinedly walked up to the front of the room and sat at the small table facing the committee. Slowly I spoke into the microphone, looking down occasionally at my notes for reference.

"This bill will directly affect my neighborhood by forcing the expansion of the Flying Cloud Landfill located next to it. This garbage dump is leaking toxins into groundwater . . ." I articulated five reasons this landfill was a poor candidate for continued dumping, and inquired whether committee members would consider living in my home two blocks from this site. Again, no volunteers.

My testimony took more than one minute, but no one asked me to stop. I solemnly returned to my seat next to my bellicose lawyer and waited for the vote. We won by one vote. We reaped a double victory that day!

We went back to Sidney's office, and once there, Sidney glared at Grant and demanded he apologize to the representative he insulted. I never saw Sidney so upset. She had daggers in her eyes. Frankly, I didn't think Grant had done anything terribly wrong. He certainly didn't mean to. After all, we knew that many representatives took campaign money from BFI in the past. Since the company never lost in the legislature, obviously many politicians supported its activities. The Senate passed the bill knowing full well the ramifications it would have on BFI's landfill. Don Storm laid out the facts well. Maybe we pricked a few consciences?

Apparently, an outburst in the legislature is uncommon. The news ricocheted throughout the capitol like a ping pong ball. I felt awful.

148

Grant's spirits fell. I thought he'd lose his job, and his reputation might be ruined forever. Josephine Nunn called Carl Jullie to tell him what happened and recommend he fire Grant. Carl politely listened to Miss Nunn, then ignored her suggestion.

Even after our extraordinary victories that day, I returned home downtrodden. That evening Dean Rebuffoni called to ask me why Grant was tossed out of the legislature. He was doing a newspaper story on the incident. I pleaded with Dean not to print it; "you'll ruin Grant's career" In a delighted voice, Dean replied, "It's a great story Susan. And besides, this incident won't destroy Grant; he's tough." That evening, I spoke several times with Dean, Grant, and Grant's wife, but nothing could undo the damage so I cried myself to sleep.

At 4:00 a.m. the next morning, the phone rang. It was Ivan calling from South Africa after a visit to the home of Hilton Davies, Chairman of the Board. Groggy, I told him the dump was ordered shut. He shared my excitement.

All the newspapers and TV stations telephoned for my reaction to the MPCA's order. For the Minneapolis *Star Tribune*, I said, "I felt like Napoleon coming through the Arc of Triumph in France as I walked out of the MPCA board room flanked by our lawyers."

To the *Eden Prairie News*, I remarked, "This shows that the MPCA finally realizes there's a problem. They're starting to listen to us. Before, we were just a couple of flies in their eyes."

I have no clue what I said on television that day. Grant said I sounded great and looked good, but I was in dreamland during the ten o'clock news that night. I was exhausted

The newspapers also reported BFI's reaction to the shut down. Dick thought BFI would comply with the MPCA order. However, if Flying Cloud wasn't allowed to expand, up to 75% of Hennepin County's garbage probably would be hauled to landfills in other area counties. I suspected this remark was aimed at other communities chosen as potential landfill sites to rally their support for the Merriam Amendment.

"Closure of the landfill is no big thing," said Robert Block, a representative from BFI headquarters in Houston, Texas. "For all intents and purposes, we haven't taken in much waste since 1986. Our emphasis for six years has been on the expansion, and that's what we'll continue to work for."

"We're going to correct the problem and get back into compliance," remarked the landfill's manager. "We were not out of compliance intentionally."

The news of our landfill victory spread quickly. Don Storm asked for a special privilege to address the full Senate. Eloquently, he informed his colleagues, "One small step for the MPCA was a giant leap for the citizens of Eden Prairie. The Flying Cloud Landfill was ordered shut down by this agency."

As I jogged in the morning along the Homeward Hills path, I met many neighbors, and we cheered our success as we greeted one another at 6:00 a.m. On a visit to Pax Christi Church, my former colleagues congratulated me on the victory. Then I telephoned CCHW and told Will's co-worker, Brian, the news of the dump shut-down. He screamed with delight and told me no one before had ever shut down an operating BFI landfill. It was a good thing I didn't know these odds, otherwise I'd never have tackled such a feat. Soon my phone rang off the hook with calls from all over the United States and Canada asking for advice on how to shut down a landfill or prevent the siting of one. I helped every person to the best of my ability and always encouraged them to press on despite opposition.

We were pleased with our accomplishment, but didn't delude ourselves into believing the road ahead was without dangerous curves. With so much money, plus their reputation at stake, BFI officials would not just lay down their swords and surrender. They'd probably retrench and come out fighting twice as hard. We were dealing with a wounded animal now. I feared the worst lay ahead. Thomas Jefferson said, "The price of freedom is eternal vigilance." Vigilant we must be until the end.

8 | LOBBYING IN THE LEGISLATURE

There was a raging battle outside the legislature and MPCA, and getting help from the community was no longer a problem. Everyone loves a winner. Everyone loves to see David slay Goliath. And with the media spotlight shining on us now, recruits were calling every day to join our ranks.

Mary Anderson was one of them. She telephoned one evening to announce in her young, pleasing voice that she had experience in writing and speaking; could I use these talents? I was delighted to have someone with invaluable communication skills and welcomed her to join us. Mary was highly motivated; she lived on Winter Place, the block closest to the dump.

Grant convinced Joe Mengel, a former ally from his Reserve Mining Days and resident of Eden Prairie, to participate in our environmental quest, too. Joe was a scientist and former professor of geology from the University of Wisconsin. In all respects, he was a giant of a man. He reminded me of Minnesota's legendary folk hero of lumber camps, Paul Bunyan; the man endowed with prodigious

151

strength, vision, humor, and cunning. I would estimate Joe's height at 6'3" plus. His shoulders and barrel chest span were massive; his size dwarfed my puny 5'7", 135-pound frame. I bet his empty shoes could probably hold several pairs of my size six models.

Besides being a towering figure of a man, Joe was a great human being. His intelligence and intuitive ability to size up people and situations helped us clearly define the problems in the case and find solutions. However, his information-gathering mode must have unnerved BFI. When Joe attended a meeting, he just sat quietly in the back of the room, peering over his half rimmed glasses at the players and took pages of notes. Later, we crunched this information and developed a strategy. As an on-the-spot mentor to Grant and me, Joe kept us sane, laughing, and on track during the last two years of the case.

After the shutdown, we focused on the legislative battle. Grant Merritt, Sidney Pauly, Don Storm, Randy Johnson, and I lobbied day and night at the State Capitol, trying to prevent the Merriam Amendment from becoming law. Finally, we won the house vote and lost in the senate. As a result, the bill was assigned to a conference committee for resolution. The conference committee was composed of a chairperson, plus five members from the House and five members from the senate. They heard testimony from interested parties, discussed the bill, and voted to break the tie. Since the amendment was sponsored by Gene Merriam, "Clean Gene" was the Chairman.

Sidney Pauly worked feverishly behind the scenes distributing information about BFI's checkered business record to her colleagues, Bob Anderson, Jean Wagenius, Dennis Ozment, Darby Nelson, and Dee Long, who were selected for the committee. However, Jean presented a special predicament. Jean's husband, Dwight, was a government lawyer and supported the MPCA's pro-expansion position. Jean, also a lawyer, mother, and activist who shut a local pornography shop, stood the middle ground before the conference committee. (Remember, she was also the one who tossed Grant Merritt out of the legislature.) Sidney sought her out for a one-on-one chat to discuss the perils of the Flying Cloud expansion.

The conference committee meeting was set up on an evening when Ivan was in town. Hurrah for small favors. I arrived early; first to occupy a front row seat and second to inform Chairman Merriam how his bill could impact our neighborhood. Up until now "Clean Gene" had refused all our phone calls and meetings, giving every imaginable excuse: too busy, late night meetings, kids sick, cat died, and so on. But this time he wasn't going to escape me. A half hour before the meeting was scheduled to begin, I sat inside the designated room, just behind the doorway, watching and waiting for Mr. Chairman like a lioness ready to pounce on her prey.

As I quietly awaited my victim, a well known lobbyist around the capitol came in and inquired who I was and why I was sitting in a room all alone. I explained. He laughed and said he'd seen Grant that afternoon wearing out shoe leather running around the capitol. "I'm glad you showed up tonight, Susan, and not Grant; you're better to look at."

Then he walked out laughing, leaving me to my solitary vigil. This brief interlude broke the tension and reminded me that once upon a dream, ten years ago, I did a little modeling. These days I was afraid to look in the mirror for fear I'd see the wicked witch of the Midwest staring back at me. At least that's how I felt after working such long hours, fighting and clawing to win this case. It was good to know that even as the undisputed "Dump Queen" of Eden Prairie, I didn't look as bad as I thought.

"Clean Gene" finally arrived on the scene, and I lunged forward extending my hand to shake his.

"Good evening, I'm Susan Varlamoff and I've been trying to contact you for weeks."

"Yes, I know, but I've been busy."

Gene diverted his eyes from my face as if I were an IRS agent coming for an audit. He walked over to the conference table to sit down and I followed behind, talking to his back about the Flying Cloud Landfill. When he turned around to take his seat, I noticed his eyes were bloodshot, no doubt from late night meetings. He seemed uneasy and cast his eyes down to avoid my stare. I continued talking and trying to establish eye contact with him. He turned his head from side to side, forcing me to do a little two-step back and forth in front of him to look into his eyes.

"Do you realize your amendment could have adverse effects on my neighborhood," I said.

"I didn't know that."

"It's been in all the newspapers."

"I haven't read the newspapers lately."

Oh sure, I thought to myself, and the cow jumped over the moon, too.

Since Gene appeared deficient in matters pertaining to our side of the case, I gave him a crash course on Flying Cloud Landfill problems. After my lecture, I gave my usual pop quiz question: "Would you live near this leaking, toxic landfill?"

Silence. Like everyone else, Mr. Merriam failed the test.

I took a seat directly opposite him in the front row of chairs. I frowned, crossed my arms and legs, and sat there giving off bad vibrations. Gradually, more people arrived and filled up half the room; BFI representatives made up the majority.

The meeting began, and every time Gene looked up, there I was with my negative body language. Learning from past mistakes, I sat through the entire evening until early next morning, so "Clean Gene" wouldn't be tempted to pass his amendment in my absence. The committee discussed the bill at length, except for the section about the landfill siting process. As the hour grew late, Gene tabled it until the following evening. Classic political trick, but this time I couldn't be fooled.

Next morning, after only a few hours sleep, I telephoned everyone I felt would help us defeat the bill. Steve Keefe, Metropolitan Council Chair, agreed to send in several representatives, including John Boy, to speak against the Merriam Amendment. The Metropolitan Council figured if Flying Cloud stayed shut and the siting process was stopped, they wouldn't have sufficient landfill space for the projected volumes of metro area garbage. "Keep the siting process alive," said a *Star Tribune* editorial.

As I recall, Hennepin County Commissioners Randy Johnson and Mark Andrews sent people over, and Eden Prairie dispatched Ric Rosow to the proceedings. Homeward Hills Homeowners Association rebels Mary Anderson, Betsy Nigon, Alex Zubar, and I went over, too, and monopolized the front seats. With only several rows of chairs in the small conference room, the pro-siting forces occupied them all.

154

There was standing room only for late comers. BFI lobbyists spent the evening leaning against the walls.

With limited testimony being taken, Ric Rosow gently argued I should be the one to speak for the city, not he.

"A statement from a landfill neighbor and mother, will go over better with committee members," he said, "than words of a hired gun."

Ric's reasoning convinced me to speak, but again I felt the horrible weight of this responsibility. One misstep and the bill could pass within hours, ending our hopes to defeat the expansion. We were down to the wire again.

I quickly wrote a few statements on my notepad and waited with my heart pounding. John Boy preceded me and presented a strong technical case for continuing the siting process. Others made convincing arguments too. I tried to give an abbreviated version of the dump's problems.

Next came a discussion between panel members. As representatives debated back and forth between themselves, we tried guessing on which side each stood. Jean seemed to be leaning our way. Finally, when it appeared everyone had spoken, there was silence. Then Bob Anderson carefully took out a speech from underneath his book and enumerated all the reasons the Flying Cloud Landfill must remain closed. (Sidney, our guardian angel, had done her job well.) During this blockbuster performance, we sat on the edge of our seats, barely breathing.

Now the vote. The Merriam amendment failed by one vote. This was the very first time in Minnesota history that BFI had been defeated in the legislature.

Mary, Betsy, and I could hardly conceal our excitement. (Alex had to leave early). Like little old ladies, we walked out of the room holding onto one another for support, giggling like school girls. After we passed through the doors, we shouted together. Then we danced through the corridors, giddy at our wonderful good fortune, and skipped down the wide Capitol steps well after midnight on a moonlit night. I don't think our feet ever touched the ground. Our spirits soared into the night. This was a mountaintop experience.

Rallying support

Besides leading the charge of the dump-buster brigade, I spoke at schools and civic organizations around town to get the word out about the dump. Anyone permitting me to address their group could have me as a speaker. We always needed to increase attendance at meetings, and with the hearing for the contested case coming up in May, we needed to pack the room once again.

Local school children were often my favorite audience. I knew they had enormous influence on their parents, since most TV commercials aim their advertising at this young segment of the population, hoping they will badger their parents into buying them the latest $100 dollar toy. And, of course, we do. I attempted the same tactic. Sell the dump's problems to the children so they would get their parents to become more involved.

Sharon Papic, wife of Bill, famous for his Billisms, invited me to talk to her third grade class. Several other classes joined hers, giving me an avid audience of 100 eight-year olds squeezed together on the floor, staring up at me. However, I had little to teach them. These children already knew from toxic leachate to volatile organic compounds about the Flying Cloud Landfill. Our media campaign worked and teachers did a wonderful job indoctrinating their young charges. The children asked many questions and had novel ideas for solving the solid waste problem. "Blast the garbage into space," suggested one boy.

Children also volunteered to help, so I advised them to write letters to MPCA officials, telling them how it felt having a leaking, toxic dump in their town. I agreed to present the letters to the MPCA Board at the upcoming meeting.

Sharon dropped off the letters at my house, and I chuckled when I read them. This is a sampling of what they said:

Dear PCA, Dear BFI,
Why did you let these people put this dump there? It is
polluting the air and it is polluting the Minnesota River.
The people don't like it, so can you please close it down.
Thanks!
Mike S.

Dear PCA,
I hope we can get that dump closed forever so it won't kill
the fish and pollute the river. I think nobody would have
thought of it, so the fish won't keep dieing and so the
animals can have clean water to drink. I am recycling pop
and beer cans!   Steve Haverly

At the middle school, Pierre's teacher asked me to discuss recycling with his sixth grade class. But my son thought the idea terribly embarrassing and dove under his seat to avoid the scrutiny of his peers. Pierre probably would have preferred to eat a plate of brussels sprouts rather than suffer such a humiliation.

On a return visit to his school, I saw an army of student volunteers pushing big carts of white paper to a central collection center. Pierre's advisor, Mr. Husaby, said they wanted to do their part to help Mrs. Varlamoff keep the landfill shut. The students adopted and implemented a school-wide recycling program. A seed planted, sprouted.

Senior citizens in town also invited me to address their group, so I visited their recreation center to discuss my favorite subject. I enjoyed them enormously. Of course, they insisted on telling me stories about Eden Prairie from the good old days, that is, the days when flowers bloomed on the prairie and the dump wasn't part of the local landscape. I listened with rapt attention to all their stories, even the shouted versions from hard of hearing members.

After concluding my tale of woe about the "garden spot of the territory" gone toxic, several spunky women volunteered to car pool their friends to meetings. I was pleased to make several more converts to the cause.

Another important caller, Mike McGowan, son of a nearby landfill owner, telephoned. He requested a breakfast meeting with me to discuss BFI's expansion. Mike explained his family also wanted to expand their landfill, but were put on hold for a litany of reasons. Funny BFI could impose its will on the government, but the McGowan family couldn't get past go with their plans. Something was awry and Mike said he had evidence to prove it. He volunteered no more information on the phone in case my line was tapped; he said he'd discuss the situation further with me at Baker's Square Restaurant.

While Mike drank coffee, I sipped hot tea, ate a big muffin, and listened attentively to his story; very interesting. He said his landfill had been issued many violations while Flying Cloud boasted a squeaky clean record, despite a history of breaking the law. He showed me documents to verify what he said. Apparently his short tempered Irish father became so enraged over the injustice he once physically threw a chair at an MPCA meeting. Mike pledged financial support and promised to testify at the upcoming MPCA meeting where the decision to grant a contested case hearing would be made. We agreed to stay in touch.

In addition to monitoring meetings and taking notes, Joe Mengel assembled a three dimensional styrofoam model of the landfill which visually demonstrated the neanderthal mentality of expanding BFI's Flying Cloud Landfill on its present site. He displayed it at our Environmental Committee meeting one evening and we all "oohed and aahed" at the result.

I wasted no time packing it up and heading downtown to the Department of Health. We hadn't heard from this agency lately. Therefore, I decided to pay my respects to the scientists and the Catholic nun just named head of the agency. The health department opposed BFI's landfill expansion several years earlier; maybe I could re-ignite its interest on the issue. Also, perhaps my strong ties with the church and years of parochial school education could get me into sister's office.

Carrying my precious model through narrow hallways, I shuffled from one overcrowded office needing repair to another. Harried officials sitting behind desks piled high with papers occupied these rooms. I finally got an opportunity to display my mini-dump to several men. After hearing my story and seeing the model, they explained they had no concrete evidence of any impropriety on the part of BFI, nor evidence of health risks at the dump. Methane leaking into the neighborhood didn't overly concern them. It was interesting that health officials would change their position on the Flying Cloud expansion. I don't recall much about my visit except I was obviously not impressing these people, so I left. Were they so overworked they couldn't take on more cases? Or had BFI influenced them?

At about this time, Grant and I decided to warn BFI's chief lobbyist, Jerry, that his client was headed down loser's lane with the expansion proposal. Grant knew Jerry from the environmentally fertile years of the early 1970's, when a bunch of guys hung out together, including Mark Andrews, to work on environmental legislation. Due to his long friendship with Jerry, Grant hoped to convince him to back off the expansion. It was a long shot, but we decided to try anyway.

Jerry proposed to invite Grant and me for breakfast at the Decathlon Athletic Club to discuss the dump. We accepted, and I arrived late because I lost my way, a common occurrence for me. (My sense of direction equals BFI's mathematical capabilities.) When I finally arrived at the table breathless, making profuse apologies, Jerry sarcastically remarked: "So you decided to show up!"

We tried some small talk to melt Jerry's glacial mood, but a frown remained frozen on his face. It broke only when he muttered some sharp comment like: "You're getting entirely too much press coverage on the case." (Tsk, tsk; how sad, I thought.)

Grant then came to the point. He looked empathetically into Jerry's eyes and said: "Jerry, I think this case will go the same way as the Reserve Mining case, in favor of the people and environment. You and BFI stand to lose a lot if we win. You're the one person who can help."

I felt good vibrations passing between these former friends. Jerry replied, less antagonistically, that BFI intended to pursue the course it set for itself and go for the expansion. Grant argued a little while longer but to no avail. Jerry didn't change his mind.

Grant left after an hour or so, and I appealed to Jerry one last time to consider our request, given the risks the landfill posed to those of us living close by. He didn't want to hear it. He made some comment about having to earn a living for his family and that I'd be running this crusade till I had grandchildren. He mocked me by imitating an old lady's voice saying, "Just one more meeting, kids."

With that cutting remark, I got up to leave. I wasn't going to be laughed at for my efforts to protect my family. I thanked Jerry for breakfast and left somewhat shattered, because he could well be right. I might be running this crusade until I had grandchildren. Maybe I was foolish to exhaust myself on a quest that might end in defeat. I drove off and went directly to Pax Christi Church to unburden my soul. Diana, the pastoral minister, listened and offered some advice: "Do something good for yourself these next few days; take care of yourself." I did, until I got over the hurt caused by Jerry's remarks.

The great monopoly game

There wasn't a moment's reprieve. April 13, 1988, the *Eden Prairie News* headlines glared: BFI BECOMES BIGGEST EP HAULER. The local owner of the biggest trash firm in Eden Prairie, Bob Carling, sold out to BFI. Bob's wife recently had a stroke and he no longer wanted to continue the business. Another case of preying on the weak to further business interests. (Remember, the land for the dump was purchased from a widow.)

I called Will to relay the bad news. This is classic garbage company practice, he said. But as always, he had an idea to combat the problem. "Run ads in the newspapers asking people to stop playing the BFI monopoly game and change haulers." What an Einstein!

I called Mary Kennedy and we sat down to design full page advertisements and have fun in the process. We came up with:

Eden Prairie
is tired of playing the
BFI MONOPOLY GAME

BFI owns: ● Flying Cloud Garbage Dump
● Trashtronics
● Incinerator for animal and hospital wastes

WHAT CAN YOU DO TO STOP THE GAME?
CHANGE GARBAGE HAULERS!

We then gave information about who to call to change garbage collectors.

Again the moms advised everyone on soccer and baseball fields, at shopping malls, and on playgrounds, about the situation, and quickly jammed BFI phone lines with thousands of calls from people switching haulers. Within months, BFI lost the trash-hauling monopoly it bought in Eden Prairie.

Our plea for a court hearing

The MPCA Board meeting to determine the need for a contested case hearing was set for April 26. Again we beat on our tomtoms to call out braves to attend this powwow. Again the city promised buses to transport us downtown, and I telephoned our new allies, Mike McGowan, Ed Crozier of the Federal Fish and Wildlife Service, and Chuck Moos, to ask them to testify. Ric Rosow, Roger Pauly, Linda Lehman and I also were scheduled to speak.

As always, Grant and I went over to the MPCA early to scope out the scene, save seats, and lobby board members. Unfortunately, the Board was working on other agenda items, so we had to mill outside the meeting room doors looking for something to do. No problem, we might be able to learn something ahead of time.

A plethora of BFI lawyers and lobbyists circulated everywhere, like worker ants in their hill. We noticed a group of them running in and out of the bathroom, so Grant joined them to see if he could learn anything across the stalls. I watched this whole affair with amusement, outside the men's room door, of course. It occurred to me that for twenty minutes I kept seeing the same men frequenting the facilities. I couldn't resist loudly proclaiming: "There must be a problem with diarrhea around here." Several men laughed heartily.

Maybe they were nervous and needed to use the restroom more than normal; so much was at stake for both sides. If the MPCA denied the contested case, we could appeal, but most likely BFI would begin digging its money-making pit in May. If we won our case, BFI would lose an enormous amount of revenue while it paid out millions of dollars in lawyers fees and expert witnesses to plead its case in court. For both sides, winning was crucial.

My BFI friend Irv came over to converse while we waited for the showdown. I teased him and all BFI's lobbyists relentlessly, but somehow connected best with Irv. I guess that's why he was often sent to pump me for information. Contrary to popular belief, I really didn't dislike BFI officials; I realized they had a job to do, however misguided it happened to be. I figured if we maintained a cordial relationship with them, they'd have a tougher time thrusting their swords into us.

Irv was short, maybe 5'5", so he sat on a table next to me to establish eye contact as we conversed. Since I hadn't seen him in awhile, perhaps he was sunning himself on the Riviera with his inflated BFI salary, I asked him how he liked our bus blockade. He chuckled and said I looked cute. (I doubt he believed a word of what he said).

Irv also inquired about Ivan's job. He heard we often moved; maybe it was time to leave again, he suggested. I replied there was no such plan. (God, forgive me for this white lie; I'll do time in purgatory for it.) Actually, Ivan's company was transferring its headquarters to Salt Lake City. Irv didn't persuade me, I think we just enjoyed our verbal exchanges.

Citizens now arrived in buses. I recognized neighbors Bill and Bonnie Swaim, Jim Gilbertson, Linda Gustafson and her baby, and many other mothers and children all walking downstairs for an afternoon outing at the MPCA. Fathers took time off from the office to join us, too. The board room was still closed, so we hung around the hallway for a good twenty minutes before we filed in and occupied the front half of the large room.

Meetings made me nervous; I couldn't sit still. As usual, I wandered around the room talking to friends and keeping my ears alert for some worthwhile gossip. I frequently changed position to shake off the BFI spy superglued to me. Sometimes, though, I'd purposely pass misinformation to someone near me to fake out our opponents. Other times, I'd just turn around and tell my secret agent to find another seat. I was tired of being followed. My only refuge was the bathroom; I rested there from time to time to take a break from the constant surveillance.

As I wandered around the room, lo and behold, I heard Dick whisper to his colleague, Lynn, "We can't fight all this opposition to the contested case; we'll have to concede." Elated, I hurried back to our lawyers to share the news. They were delighted to know we won the battle for the contested case before the meeting began.

I continued working the room, chatting with Ed Crozier, Mike McGowan, Chuck Moos, Sidney Pauly, Patricia Madame, Mary Kennedy, and many others. The mood was lively. The mothers all thought Mike looked handsome in his suit, and we all thought his proposal to accept an enlargement at his landfill as an alternate plan to the Flying Cloud expansion was an excellent idea. Ed Crozier and Chuck Moos gave impressive speeches discussing the impact of the expanded landfill on the refuge. Linda showed the nine member board a U.S. Geological Survey map made years ago at the request of the government, defining those areas suitable for landfills. The Flying Cloud Landfill site was marked in red, indicating the worst possible location for a garbage dump, except perhaps for Lake Minnetonka.

Roger spoke next. He took on the issue of BFI's suitability. That was an easy topic, but lengthy. Roger gathered steam as he read down the list of cases and violations against BFI. By the end of his talk, he couldn't contain his rage anymore. Shouting, he quoted a Boston BFI manager giving advice to his subordinate about the competition, "Put Kelley out of business, do whatever it takes. Squish him like a bug." Then Roger remarked, "Is this the kind of corporate citizen we want in Minnesota?"

Roger lost track of time; easy to do when you get on a subject like BFI's corporate character. Only 30 minutes were given to the city and its citizens; our time was nearly up before I even got a chance to speak. With only a few minutes remaining, I tried to wrap up our presentation. "A contested case hearing will show you very clearly that Flying Cloud Landfill is an unsuitable site run by Browning Ferris Industries, an unsuitable company. Recommending an expansion would be an unsuitable decision."

Dick followed us to the podium. He disputed the attack against BFI's reputation nationwide, and described the "exemplary" records of both BFI and Woodlake within Minnesota. "I apologize I didn't wear my black hat, my black suit, and ride in on my black horse," said Dick. "I'm not colored for the story you just heard."

We laughed so hard, we nearly fell out of our seats. Dick was a regular comedian these days. Maybe *Saturday Night Live* could use his talent. He finished by saying, "I want to encourage you to have a contested case hearing so we can get at the facts."

Well that jived with what I overheard him say to Lynn. Of course, prior to this meeting, BFI was adamantly opposed to a contested case hearing. Dick said in the newspapers, "We feel the concerns have all been addressed, and our permit application contains the information showing the site can be operated according to the rules." Roger Pauly suggested Dick took this stand to avoid the appearance of fighting a losing a battle.

At this meeting, Eden Prairie also requested MPCA to examine BFI's record as one of the contested case issues. Nationwide, BFI was an industry leader in violations, while its record in Minnesota was nearly spotless. City officials cried foul. They wanted to solve this riddle through the contested case. Since MPCA regulations required the permittee be "fit and able," it seemed fair we find out if Browning Ferris Industries could meet these criteria. There was one problem: if we dug at the corporate record of BFI, city lawyers might be at personal risk, considering the rumblings we heard about the company. Also, uncovering Woodlake's record in Minnesota could prove an embarrassment to the State.

Van Ellig, a young, good-looking board member and lawyer, argued vehemently for us. "Clean air is clean air, be it in New Jersey, Ohio, or Minnesota. If they (BFI) have violated federal clean-air standards in other states, or not, I want to know about it." We were awarded the contested case, but the Board didn't yield on the fitness issue. The meeting was a mixed victory.

We next worried about which judge would be appointed to the hearing. Again the stars were with us. Allen Klein, a judge of integrity and respect in the Twin Cities, was chosen to preside over the case. Now months of preparation loomed ahead. Evidence had to be secured, witnesses chosen, depositions taken, testimony prepared, and dates set, but at least we had an opportunity to present our facts before an impartial judge.

BFI's expansion case wasn't looking brilliant now. It's recent losses were taking a heavy toll on the combatants. Very sadly, we learned the landfill manager had a fairly severe heart attack. I felt bad he was sick. He was always respectful to me and agreed that if he lived where I did, he'd fight the expansion, too. He continued coming to meetings, but was very tired. He called me Susie, the name my father called me as a child, and smiled down at me with understanding. I returned compassion. I couldn't help thinking he must have a daughter my age.

164

Dick now quit the case and changed law firms. We heard through the rumor mill he said he got tired being on the wrong side. Contrary to his remarks, he must have believed black wasn't his color. Irv, or "Shortee," quit lobbying for BFI. I missed bantering with him.

There was a big shake-up at BFI. The district and midwest regional managers were fired. Several other executives were given pink slips, too. However, the company hired a public relations person for damage control. This was an impossible job, since company problems ran so deep. Ironically, Tom Vandervort, who took this position, shared a friendship with Jack Bolger, who was our acquaintance, too. Poor Jack became the monkey in the middle as Tom and I questioned him. I finally told Jack to relay this message to his friend: "You will lose this case, so save face and get out."

165

9 | AU REVOIR EDEN PRAIRIE

M ay is a beautiful time of year in Minnesota. As earth greens with new life, spring fever hits and local folk crawl out of homes into blinding sunlight after months of hibernation. They hose off the last remnants of snow from their lawns and linger in the streets to exchange last winter's gossip. I always found myself impatient to see our garden's annual flower show, so I often dug down into the rich, black soil looking for tulip heads ready to poke through the surface. Spring in Russia is much the same, I'm told; a rush of new life.

While lawyers searched for expert witnesses to testify at the contested case hearing, citizens brainstormed for some creative activity, preferably something outdoors in this warmer weather, to demonstrate our spirits hadn't frozen over the winter. A member of the Environment Committee noticed a letter to the editor written by Woodlake Sanitary Service's (alias BFI) new district manager announcing: "We would be happy to welcome anyone to tour our facility at any time in the future." That was it! A field trip to the dump!

Since "any time" was fine, we concluded reservations weren't necessary, so a dozen of us planned a surprise spring fling to the landfill. We found this idea amusing and relished the preparation. A good mix of men, women, and children would be good, we decided. Also, we resolved to carry every camera we owned to appear like tourists. Having this equipment handy would also enable us to photograph some irregularity in BFI's operation during our sightseeing trip. We selected the date and time; we all knew the place.

The Swaims, Maureen Wilson, my mother-in-law, Louis, myself, and five or six others arrived at the scheduled time dressed in our jeans, Reeboks, and light sweaters (a cool breeze blew off the bluff this time of year.) Looking like Japanese tourists at Disney World, we carried an array of camcorders, tripods, and cameras as we walked up the front entrance to the Flying Cloud Landfill office.

Once inside, we straightened up and asked for the advertised tour. I brought along the newspaper clipping to show what we meant. The secretary appeared uneasy and asked us to sit in the lobby while she checked for clearance. In a muffled voice, she made several fast phone calls in succession. Within minutes, a truck, kicking up dust, roared in our direction and screeched to a halt just outside the office. The landfill manager came barreling in.

I walked over to him and inquired about the landfill tour. Again, I presented the newspaper piece. He went into the office to make more phone calls and asked us to wait again. After twenty minutes or so, we were told a tour wasn't available this day; heavy construction was underway, posing a danger to us. Besides, we should have called ahead to arrange for a site visit, the manager said.

I was impressed with this alibi. It was more sophisticated than the previous ones. The manager finished by explaining we'd have an opportunity to see the facility soon, and would be notified of the time and date. For now, we had to leave the premises.

We argued a bit about the deception we felt at being denied a dump tour. False advertising, we charged. After we made our point, we departed with our cameras, confident BFI realized we hadn't taken a vacation from our activist activities.

Soon after this aborted landfill tour, Bill Swaim came running over to the house one day, waving an article from *USA TODAY*. It stated the EPA was proposing tougher standards for landfills by restricting them near airports, flood plains, wetlands, and earthquake zones. Minus the earthquake zone, we failed the new criteria on all counts.

Public hearings for the new regulations were being held in all major cities except Minneapolis. Why this oversight?

Since we knew the MPCA must eventually enforce these federal standards, we figured success was in the bag now. Grant, however, figured differently. Having worked in government, he cautioned us that bureaucracy moves at a snail's pace, and it could be years before these standards are "promulgated." (Impressive government jargon for became law.) But if you want to try, go for it, Grant said. Just don't expect much. We should have listened to our lawyer.

Excited with this seemingly wide open avenue, we telephoned EPA's regional office and spoke with several bureaucrats. One empathetic woman wrote a few letters and discussed the case with her superiors. According to her, the Flying Cloud Landfill case made its way to the top of the agency. However, after several months of phone calling and letter writing, we were told the state agencies have the responsibility to enforce Federal regulations. If we believed they were shirking their responsibilities, then we could sue.

"Yes, but how can average citizens protect themselves from a dangerously polluting situation if they don't have a spare million dollars for lawyers' fees?" I asked.

"If there are hazardous waste violations, the Federal government could step in, but there must be clear evidence to support this charge," we were told.

"Suppose you don't own a backhoe to dig up toxic barrels on a site for which you may not even have access."

There had to be strong evidence of leaking chemicals, came back the answer.

We tried and tried, but got nowhere; we played a virtual "ring around the rosy" with the EPA and all fell down. It seemed dead bodies were needed in the streets before the EPA would investigate. One Love Canal was too many; we refused to let conditions in Eden Prairie get that far out of control so we were on our own again.

Then it struck us that BFI's CEO Bill Ruckelshaus probably still had friends at the EPA from his days there as chief. We also heard through the CCHW grapevine that Ruckleshaus advised President Bush to appoint his friend William Reilly to head the EPA. His suggestion was taken. After thinking the situation through, we now understood why we got no satisfaction from the highest environmental pollution control agency in the land.

Scientific support

During one of my many conversations with Will, I asked if he had any data showing health and safety problems at a landfill similar to ours. Will put me through to Steve Lester, technical director of the organization. Steve said, very few studies exist, because problems are just surfacing in neighborhoods adjacent to mismanaged garbage dumps. Even at Love Canal, scientists still don't technically understand why there was such a high incidence of illness and death.

Risk assessment is the government's method of evaluating health effects from individual toxic sites. This effort is based on exposure limits the EPA sets for individual compounds, using a 160 pound healthy male as a standard. Of 35,000 chemicals on the market considered definitely or potentially harmful by the EPA to human health and safety, guidelines have been established for roughly 2,000. Meanwhile 1,000 new ones are added to the global chemical inventory each year. Not only is the EPA unsure of types and quantities of toxic waste produced by industry, but little is known about effects of individual toxic chemicals on the environment, and nothing on how they act in combination.

Since landfills contain mixtures of many chemical substances, it remains nearly impossible to know how a concoction of compounds affects a person. Since weight, age, size, and general health of each human being varies, other unknowns are added. The temperature, wind, distance from hazardous substances, and length of exposure must be factored into the equation also. The result: no one can identify precisely why citizens living near landfills are getting sick, but newspapers across the country and around the world are reporting this phenomenon. And if risk assessment didn't work at Love Canal, a worst case scenario, then there was no hope this model could successfully determine health risks posed by BFI's Flying Cloud Landfill.

At one unlined dump site of sand and gravel in Seattle, Washington, methane gas leaked half a mile into adjacent homes, Steve told me, forcing evacuation of eleven families. He sent me the newspaper clippings. This situation paralleled ours closely. I hoped it might scare some sense into Minnesota's politicians. Relocating hundreds of families is expensive, and bad press generated by gases leaking into homes would prove embarrassing for city, state, and federal pollution control officials.

I xeroxed the methane gas migration article, passed it around to our group, and inserted it into city council member's folders. To my more valued government friends, I hand-delivered it. Metropolitan Council Chair, Steve Keefe, was first to receive the information. Unfortunately, I arrived late for my appointment, probably because I absent-mindedly drove the wrong way down a one way street and got lost again. When I rushed into his office, Steve was tenderly nursing a sore foot. After I commiserated with him on his misfortune, I quietly presented him with the Seattle story about methane migration and the resultant catastrophe. Also, I gave him news clippings with pictures to confirm my story.

Steve showed no interest. He proclaimed his decision to recommend expansion of the Flying Cloud Landfill was a good one. I was depressed. This man must have a heart of stone, I thought. Worse yet, I wasted an afternoon away from my children, who were now home from school for summer vacation. It took me a few days to recover from my misery.

A week or so later, I packed up my papers and headed downtown again, a real glutton for punishment, to Hennepin County Commissioner Mark Andrew's plush office. My reception was better. Despite feeling sick, Mark was in good humor and attentive as always to what I had to say. Together, we went down to the cafeteria for something to drink, and I repeated my Seattle methane migration saga, showing him the article. Mark took in everything and seemed concerned. We returned to his office and discussed the Twin Cities solid waste problem in general and I said, "Mark, there won't be an expansion of the Flying Cloud Landfill, and since you don't have much landfill space left at other sites for the county's garbage, there really needs to be mandatory recycling in Hennepin County."

"Yes, I know, but I'm afraid it will be difficult to enforce such a regulation."

"In Japan, they solved that problem very simply. A person's garbage just isn't picked up if recyclable items are not pulled out."

Deep in thought, Mark nodded in agreement.

We both studied how the Japanese handled solid waste, since they solved the problem long ahead of most developed countries. Why? Because they ran out of landfill space years ago on their tiny islands. Back in the 1960's, angry citizens took to the streets protesting polluted drinking water caused by leaking dumps. As a result, Japan instituted tough recycling and composting programs. Garbage that's neither recycled nor composted is incinerated in furnaces outfitted with the most advanced pollution control equipment to generate electricity.

Mark had half the solution in place with the incinerator downtown, but something needed to be done about closing the county dump and mandating a tough recycling program. Chemistry between us was good that day. Mutual trust and common concerns for the Twin Cities' solid waste problem, which had a dramatic impact on both our lives, encouraged us to search for solutions.

Within weeks, the Minneapolis *Star Tribune* reported that the county board, spearheaded by Mark Andrews, voted to begin a mandatory recycling program in Hennepin County. My instincts about this man were good; he had vision and the will to carry it out. I was pleased to be part of the solution and not just the problem. I applauded Mark's "aggressive handling of the solid waste problem" in a letter to the editor.

I saw Mark once again that summer. Between lobbying officials downtown, I planned some family outings with the boys. We had never been to the annual State Fair, so we set off early one morning in a Herculean effort to cover it all in one day. We ate our way from the farm equipment to the 4-H winners, stuffing ourselves on such fair fare as hot dogs, french fries, coke, and cotton candy. There, wedged among the tractors and the combines, was Mark Andrew's famous french fry stand. And there he was, Twin Cities' political powerhouse working alongside farmers and their wives at this popular event. Mark wasn't dumb. What a way for a politician to meet people!

My boys and I walked over to purchase fries and surprised him. He seemed happy to see us. I inquired about the business and Mark was pleased to give us its full history.

"We had a bad beginning. The first year the stand collapsed; the roof caved in." Then another year, we ran out of potatoes. I put out a bipartisan SOS for more and quickly got them, thank goodness."

Mark further explained he opened the stand at the fair during his college days to raise money for tuition. Now he did it for fun and an IRA fund. It impressed me he persisted, despite initial setbacks. Someone less tenacious might have given up. I often felt perseverance made the difference between failure and success. Now I felt sure this was true, because Mark was a winner.

A dump tour

Eventually BFI fulfilled its promise to invite landfill neighbors to an open house to show off a newly installed methane extraction system. For that afternoon frolic through the dump site, we doubled our previous invitation list and surprised BFI with engineers, politicians, and interested friends. The first part of our tour was held in a conference room, where officials served us soda pop. Seeing an opportunity to have a little fun, we asked whether the carbonation in the beverage came from the landfill gas. All two dozen of us broke out laughing. However, Tom Vandervort, BFI's newly hired public relations person, didn't share our pleasantry, and more or less told me to shut up. With our series of wins, we felt brash and openly joked around. As Grant said, "We had them on the ropes, now."

Using colored charts and maps, landfill officials explained the methane extraction system, pointing out its many features Same old song. Afterwards we went outside to tour BFI's landfill. The manager drove us right down into the pit. We held onto the seats as we plunged down, like we were riding a roller coaster. Following close behind us were several neighbors' cars.

At one point along the tour, John Kennedy jumped out of his car to investigate steam rising from passive methane vents. Worried for his safety, Jerri Coller ordered him back into the vehicle. When he returned, everyone collectively said, "John, you stink. Open the windows." This incident indicated there were other smelly gases mixed in with the odorless methane.

We got out of our vehicles to inspect the methane extraction pipes in open trenches. Grant, Bill Papic, and I were deep in discussion when we saw BFI lobbyist, Jerry, standing behind us, leaning in our direction and listening to our conversation. I turned around and playfully told him we didn't want him spying on us, imitating his incline figure.

The one positive aspect of the tour was visual assurance that the methane extraction system was installed to control landfill gas migration.

Now the dump required constant monitoring to assure the system worked properly. This landfill would probably emit methane for many dozens of years into the future. Would BFI be around that long to assure the safety of its system?

On the move again

This was our last summer in Eden Prairie. Ivan's company announced it was moving its headquarters from the Twin Cities to Salt Lake City. With high Minnesota taxes and pro-union laws, the mining company found it difficult to show a profit, especially when the industry was in the doldrums world wide. This came as a big blow for employees, since the company was started in Minnesota by the Longyear family one hundred years ago.

Ivan was offered a transfer to North Bay, Canada, where the company decided to headquarter its mining operation. Originally, Ivan accepted the position. But on visiting the area, we couldn't imagine ourselves in this northern outpost of Canada, four hours from the nearest city, freezing through eight months of winter. A native Canadian told us, "The lakes break up in May."

Finding other employment in the Twin Cities proved difficult. Many companies had slashed middle management jobs, resulting in a large pool of highly qualified management people pounding the pavements looking for work. Ivan also decided to change fields, since the future for mining looked bleak. However, without experience, no one was eager to hire him when so many others were available requiring no training. Times like these try the strength of families.

Eden Prairie was the only place we ever considered home. In our 14-year marriage, we moved ten times around the world. Pierre was born in Tehran, Iran, went to preschool in Belgium and Phoenix, competed on a swim team in South Carolina, and played ice hockey in Minnesota. We spent seven and half years in Eden Prairie, had wonderful friends and neighbors, enjoyed inspiring church services, great sports teams, excellent schools, beautiful countryside, and lasting memories. Yet, it was the strength of our community, the spirit of neighbor helping neighbor born in Eden Prairie's pioneer days, that most impressed us.

After an overnight blizzard dumped two feet of snow on the ground, I often awoke to the sound of John Kennedy's snowplow, digging us out at 6:00 a.m. before he did his own driveway. Shoveling our driveway was a difficult task with three toddlers and a husband out of town. When I had emergency surgery, the church and the neighbors brought meals to our family for two weeks. In Bluff's West, several mothers and I formed a baby sitting co-op to get through those early years at home with young children.

Year after year, Ivan, a Belgian, who played soccer during his youth, was a welcome presence on the soccer fields donating his talents to coach the boys' teams. At Pax Christi Church, I volunteered my time to set up and run the children's religious education program when my boys participated in it.

Finally, it was in Eden Prairie, Minnesota, that my neighbors and I walked arm and arm down a long, exhausting path to shut a leaking, toxic dump threatening our families' health and safety, and the future of a magnificent wildlife refuge.

How do you leave a community where you've devoted such effort and have received so much love? How does one tear out family roots that run so deep? How does one leave a half completed crusade that could reverse, jeopardizing the lives of friends and setting a dangerous precedent for others? The answer: with great difficulty, much pain, and rivers of tears.

Ivan and I agonized over the decision. After all, crusading doesn't pay bills and feed children; it only nourishes the soul. Since I hadn't worked outside of the home in years, I couldn't earn the equivalent of Ivan's salary while he searched for work. Employment prospects were slim anyway, due to massive layoffs. If I did find a job, then my dump activities would cease. Would there be someone else to take my place? Or would we be stuck next to an expanded landfill, and perhaps suffer dire consequences? Though I intellectually realized moving was the right thing to do, my heart still ached at the thought of separating from all we cared for. We had no choice but to accept the company's alternate offer of work in Pennsylvania.

Vacationing In northern Minnesota

With the impending move, we decided to take our last summer vacation in Northern Minnesota. Grant raved about the beauty of Duluth and the Boundary Waters, and insisted we see it before we left. Deep in

175

my heart, I couldn't imagine Duluth, Minnesota, was the splendor he
described. After all, I'd been lucky enough to visit some of the great
cities in Europe and secretly thought my Minnesotan born and bred
friend was overstating the facts. But we took his advice and packed up
our same blue station wagon, now ten years old, with our immediate
family and Grandmaman, then drove four hours north to Duluth on Lake
Superior.

The land was everything Grant had described and more. Duluth sits
on a huge scenic bluff that gently slopes to Lake Superior, reminding me
of Italian coastal bluffs dropping to the Mediterranean Sea. And when
you look out across the lake on a clear day, as the song says, "You can
see forever."

We stayed a few days in a hotel on Lake Superior to give the boys
hour upon hour of carefree living along the rocky shoreline. They
skipped pebbles on the lake and scampered up and over boulders like
rabbits. One morning while also vacationing in the area, Grant and his
wife met us for breakfast on the hotel's deck facing the water.

What a marvelous day! Grant and I walked to the deck's edge to get
an unobstructed view of this vast lake. The air warmed us ever so
slightly as we gazed across the water sparkling in the early morning
sunshine. It was so big, one of the world's largest lakes. It reminded me
of the sea, and like the sea, waves broke rhythmically on the shoreline.
The noise was so quieting, it soothed the spirit. This was a tender
moment; the kind one tucks away to recall during trying times. Now I
understood why Grant worked so hard to stop companies from using this
lake as a sewer.

After our short stay, we crossed the state and national border into
Canada to see a restored fur trading post and a replica of an Indian camp.
The boys enjoyed reliving adventures of voyageurs from "olden days,"
canoeing along a river singing songs, bartering for furs, and exploring
trappers' primitive cabins.

Later, we visited a nearby Indian village to learn about America's
indigenous people. "When we kill an animal, we use every part of it,"
said our native American guide. "The flesh is eaten for food, the hide
used for clothes, and we carve the bones for fish hooks."

"Indians are good environmentalists," I added. "They understand
their relationship with the earth. They only took what they needed for

survival, leaving the land, air and water pure for future generations."

I was sorry to have injected my opinion. The Indian became enraged. He started yelling at me. "You white people have destroyed the land for everyone in these last hundred years. We lived on it for thousands of years without harming anything." I felt terrible. He was absolutely right. It made me ashamed. I begged him to listen to me.

"I'm on your side." I pleaded. I'm fighting for clean water and land. My friends and I shut a leaking landfill," I pleaded. The Indian didn't seem convinced; he stayed angry and continued shouting at me. Somehow I got away, but it hurt to hear the truth from a man whose culture held the earth sacred for millenniums, only to have it ruined by us. White men who landed on the shores of our continent described these red-skinned people as savages, but I believe it is we who merit that name.

In my environmental work, I studied the culture of the Indians and admired their understanding of nature and relationship with it. Often I ended my speeches with the eloquent words of Chief Seattle who spoke on this theme when he sold the white man land in 1854.

Every part of this earth is sacred....
every single pine needle, every shore,
every mist in the dark woods, every
clearing, every humming insect is holy
in the memory and experience of our race.

You are part of the earth and the earth
is part of you.
You did not weave the web of life,
you are merely a strand of it.

Whatever you do to the web, you do to yourself.
You may think you own the land;
you do not....It is God's.
Love the land as those who have gone
before you have loved it.
Care for it as they have cared for it.
Hold in your mind the memory of the land

as it is when you take it.
And with all your strength, with all your mind,
with all your heart preserve it for your children
and love it as God loves us all.

Dump headquarters at the Residence Inn

Mentally, I put off our move from Eden Prairie as long as I dared, but by July, I had to face reality. I called Patricia Madame and her husband, Duane, both realtors, to prepare papers to sell our house. When the For Sale sign went up on our front lawn, I wept. Every time I drove in and out of the driveway, I cried again upon seeing the placard. There was no denying it, in a few months Eden Prairie would no longer be home.

The initial shock of moving exhausted me. One day, when I joined friends for the annual City picnic at Cummings Homestead, a pioneer farm, I felt like a rag doll. My arms and legs couldn't hold up my body; I had no energy. Seeing my state of being, Patricia Madame took me by the hand, like a mother, and walked around leading me. The child in me gratefully accepted the guidance.

At this community affair, politicians were encouraged to give short speeches on a soap box, reminiscent of early political days in Eden Prairie. After Patricia Madame gave her spiel and stepped down, she pushed me up to give my anti-dump talk, so I did. It was an enjoyable experience and another small respite in my current career as an environmental activist.

Incredibly enough, our house sold in three months. But only after we lost three offers. A garbage dump within walking distance isn't a big selling point, we discovered. In October, 1988, a van pulled up to our home to take our household belongings into storage; the new house we were building in Hummelstown, Pennsylvania, wasn't yet completed. In the meantime, we decided to stay at the Residence Inn in Eden Prairie. I wanted to postpone our departure for as long as I could, savoring a few more months of life on the prairie, and leading the landfill quest until the last possible moment.

Pierre, Neil, Paul, and I moved into the Residence Inn with some of our clothes and all my dump files. (Ivan took an interim position in Canada.) HHHA headquarters was relocated here until I left. We had our

178

weekly meetings in the Inn's lounge, where the wonderful staff served us refreshments, and switchboard operators kept busy fielding our calls and taking phone messages. City school buses made a special detour out to the hotel to pick up the children for me. "We don't normally do this," said a school official, "but for all the work you've done for Eden Prairie, we want to make this exception for you."

The contested case hearing nears

Preparations for the contested case were well underway now. After haggling with MPCA staffers, four issues to be addressed in the case were the following:

1. Whether BFI's expansion would significantly exacerbate existing contamination.
2. Whether the liner and leachate detection system specifications were adequate to prevent leachate from migrating to ground water.
3. Whether methane gas migration could be controlled adequately to prevent gas from posing a safety threat to neighboring residences; and
4. Whether an expansion of the landfill would significantly increase the potential for failure of the bluff

Judge Klein held a series of pretrial hearings to establish certain parameters in the case: who bore the burden of proof, who could testify and when, ground rules for testimony, and so on. Since there were many such meetings, I can't recall details of each, but I do remember blurting out my concerns on health and safety issues. Legal protocol demanded that all comments be addressed through attorneys, but I really didn't care if I violated it. Too much was at stake.

In one such instance, I vividly recollect Judge Klein asking who should have the burden of proof. Naturally, Dwight Wagenius teamed up with the MPCA, pointed to us, and said, "Them." I sat terrified. There was no way we could prove the landfill was a health hazard without dead bodies in the street, and I also knew BFI couldn't demonstrate conclusively a landfill expansion was safe for neighbors. Grant and Ric made no comment; they seemed caught off guard.

179

"Didn't the MPCA shut the landfill," I observed, " because methane gas was migrating off the site and they felt it was a hazard to the neighborhood."

"Oh yeah," said Dwight, "BFI should have the burden of proof."

Again within seconds, the battle to stop BFI's expansion was nearly lost. It took an hour for my heart to return to a normal pulse after that shock.

During pretrial hearings, it was decided a limited number of citizens could testify as witnesses in court. Two open meetings were set in Eden Prairie, where the public could comment on the issues. The contested case hearing was scheduled to begin in November, and BFI witnesses would testify first.

Our work was cut out for us. I had to find citizens who knew the case well enough to act as semi-experts. Next, Grant Merritt had to screen the candidates. After that, Chris, BFI's trial lawyer would depose the chosen few. Finally, they would be scheduled to testify. We frantically exchanged information and ideas in our neighborhood as fifteen of us prepared to face Chris.

In the selected group were Sever Peterson, area farmer; Tim Homes, a neighbor with an unobstructed view of the dump, Bill Swaim, neighbor and biologist; Scott Anderson, a police officer living in a dumpside home; Mary Anderson, a writer and speaker sharing Scott's abode; Alex Zubar, a concerned citizen living near the dump; Cary Cooper, an accountant and former dump leader; Stan Johannes, a banker and former EPCC president; Pete Sadowski, a biochemist and former EPCC leader; Mary Kennedy, a mother and HHHA leader; Lori Kjseth, a Winter Place resident; Kathy Palmer, another Winter Place resident and early activist; Andrew Detroi, a Metropolitan Airport Commission employee and feisty Hungarian; and lastly, me. I have included all these names because it took courage for these people to speak out for our group when they could have remained silent.

I asked to testify last because I remembered Warner's warning, "Chris tries to topple the lead domino to make the others fall down behind it." I didn't want others to lose faith if I caved in under the pressure.

We also had to determine dates and places for the two open hearings in Eden Prairie, and get out flyers to the public, telling them about this opportunity.

My days in Eden Prairie were numbered now. However, I had not yet spoken with Governor Perpich, so I drove to the Minnesota State Capitol and once again presented my reasons why BFI's Flying Cloud dump must not be expanded. The Governor's top aide acted unimpressed, and made some comments about every neighborhood having problems. "Some people complain about noise pollution," he said, " others have wells contaminated by nitrates, and so forth." I was angry. I stressed that our situation was imminently dangerous and the MPCA closed the landfill due to methane migration.

"We want to meet with the Governor soon," I declared.

Making no promises, the aide escorted me out of the office verbally patting me on the head. I assured myself I wouldn't rest until the citizens met with Rudy Perpich, Governor of Minnesota.

The action was nonstop now. I crammed every waking minute with meetings, speeches, and lobbying. During this period, Sidney called very excited to tell me she was able to book me as a speaker for Rotary. I accepted. With so much going on, I forgot to bring my speech the morning of the meeting. I raced back home to get it, barely arriving on time for my talk.

Sidney gave me a wonderful introduction describing me as that nearly extinct species of human being, the "stay-at-home" mom. The audience included some neighbors and Jean Anderson, one of the few women in the group. The Mayor and City Manager were no shows.

The group listened attentively and even laughed at my jokes. (My children nearly convinced me that I was incapable of making people laugh with my stories.) When I was through, several people came over to speak, so I stayed as long as necessary to speak with anyone who cared to listen. It was vital we reach our city's business leaders, because if they stood behind us, our effort would succeed.

By late fall, just before my departure, BFI adopted the strategy of trying to discredit me as head of the Association. Dr. Warner warned me. Chris summoned me to court three times for this purpose. One such occasion was scheduled on Halloween afternoon. BFI officials were well aware I cared for my children at the hotel alone, and obviously the only parent who could take our sons around trick-or-treating. Chris kept me downtown late this evening, cross-examining me about our organization.

He demanded a list of our Association's contributors and members. I flatly refused. The last thing we needed was BFI blackmailing our patrons.

Finally, Judge Klein told Chris to lay off this line of questioning. It was past six o'clock when I pulled into the Residence Inn parking lot. Standing outside on the sidewalk in their costumes, were my three sons, waving at me and calling out. They hadn't eaten dinner yet. I felt terrible. But they were resourceful. They made arrangements to have the father of Neil's friend pick them up to go trick-or-treating in the old neighborhood one last time. Barely had I pulled in when the boy's father whisked my sons off. I cried. First, because I wanted to go with them myself, and second, because I felt so fortunate to have such incredible support. There was no sit-down dinner for the Varlamoffs that evening. Snickers and M & Ms on the run would be supper for my boys tonight.

In addition to an outpouring of help, I was offered political positions. Several of our financial benefactors proposed I run for City Council or Mayor. They would fund my campaign. And a representative from the Metropolitan Council invited me to be a candidate for the Council.

Then the *Eden Prairie News* requested an interview with me for a feature story before I moved. For the last time, I appeared on the front page under the headline, "Leader of landfill opposition soon to leave Eden Prairie." All of this hoopla didn't make it easier for me to leave, but, for now, I worked 14-hour days not giving a thought to my imminent departure.

Wanted: a leader for the quest

One big problem concerned me now: there was no one to command the dump buster brigade after my departure. Seven years of time, money, and energy had been invested into this crusade, and so much had been accomplished. BFI's landfill was shut, a contested case hearing was granted, BFI was beaten back in the legislature, the hauling monopoly it acquired in our city was gone, we amassed enormous support, and community fervor was rising to a roar. The opportunity to make environmental history seemed close at hand, but without a committed leader at the helm of the organization, all could be lost.

At this point, no one person wanted to dedicate the long hours necessary to shoulder this responsibility. Finally, I believe it was Grant who came up with the idea of staging a rally, where Lois Gibbs could tell her horror story of the Love Canal, and motivate people to grab the baton and sprint the last mile to the finish line.

I telephoned Will and inquired whether he and Lois would come in December to relate the Love Canal story and encourage citizens to take charge. Will said Lois just had a major operation, but felt she would be recovered enough to come. He would ask her. Luckily, Lois accepted our invitation.

Putting together a massive community event required a good organizer and we were lucky Duane Pidcock, husband of City Councilwoman Patricia, volunteered for the job. He had an excellent track record around town. Duane spearheaded an annual ball that's now a major event in Eden Prairie, raising thousands of dollars for charity. He also served as manager of his wife's and friend's successful bids for the City Council. Duane Pidcock knew how to successfully move the town's shakers.

Since Duane took the lead, again I followed the leader. Early on in the crusade, I learned to find the best people for a job and give them the resources they needed to do it well. So Duane joined our Environmental Committee meetings at the Residence Inn and proposed his strategy. His style prompted me to nickname him Adolf, because he had us clicking our heels and goose-stepping to his orders, as he assigned us tasks. Duane also checked on us to make sure we executed his commands and weren't drinking beer at the local bar and having fun.

"We need a slogan. Then we'll repeat it over and over all around town," instructed Duane. "How about, 'Lois Gibbs is coming.' This statement will arouse curiosity in people less astute in environmental catastrophes than us, prompting them to find out who this woman is and what she's all about. We'll paper the town with green flyers giving a short biographical blurb on Lois, and the date, time, and place of the rally. Then every bulletin board and marquee in town will carry the news, 'Lois Gibbs is coming.' We'll rent a huge portable sign to display the slogan and put it at the city's busiest intersection. And Lois's arrival will be trumpeted in the churches' and schools' bulletins and newsletters. We'll assemble press packets and hand deliver them to newspapers and TV stations, and a press conference will be scheduled at the airport." All of the above was accomplished on time, thanks to Adolf.

183

It was a wonderful community effort. A contingent of 45 Eden Prairie citizens volunteered to pick up flyers at the Residence Inn and get them out to their neighborhoods. Louis towed a giant portable billboard down Flying Cloud Drive with his family car, and parked it in front of the dump. Duane arranged newspaper advertisements; he knew what section and pages drew the heaviest readership (first section, third page). A local businessman printed flyers at his expense. Father Tim turned over the church to us for the evening. The Residence Inn offered to lodge our guests free, and Flagship Athletic Club offered to host a breakfast the following morning for Lois and Will and the Environmental Committee. Several friends across town cooked wild rice soup for our dinner. Lori and several other Environmental Committee members offered to set up a donation table the night of the rally. A guitar group agreed to sing for the event, and my Mother flew up from New Jersey.

This would be my last public appearance as a resident of Eden Prairie. For everyone's morale, principally my own, I decided not to mention this at the rally. We all agreed I'd fly back to testify in court and assist my friends in anyway I could, so I wasn't leaving forever. But it wouldn't be the same and we all knew it.

I wanted to look my best for this night. I searched the shopping malls for just the right outfit. At the Laura Ashley shop downtown, I purchased an emerald green, drop waisted dress with a sailor collar trimmed in black. The rich green color represented the earth, and the sailor look stood for the sea. I arranged with Marsha, our neighborhood Mary Kay representative and wife of Pete, our biochemist, to purchase green eye shadow and soft pink lipstick to complete my environmental attire.

All systems were "Go!"

Suddenly, on the night before Lois's arrival, I got a call saying the big portable billboard positioned in front of the dump was missing. What could I do so late at night? Since Louis installed it, I called him to find out what happened. We yelled back and forth, made a few phone calls, and learned that a BFI official had it removed because he calculated it stood a few inches on the county right-of-way. I thought it was a typical, underhanded shot to shake my confidence for the next day's events.

After only a few hours sleep, I awoke before the family at 5:30a.m. to work on my speech. I tiptoed in the dark to the bathroom, where I wrote it sitting on the floor. That afternoon I drove out to the airport with Betsy Nigon, a neighbor who joined the group recently. We picked up Lois and Will while my mother watched the boys. I had only seen Lois in photos; I hoped I'd recognize her.

Lois Gibbs flies to the Twin Cities

I spotted Lois as soon as she came off the plane. She resembled her picture and stood a thin 5'4." I figured the guy next to her was Will. We embraced. It was good to see them in the flesh, to look into their eyes, touch their skin, and hear them laugh.

We walked to the airport room for the press conference, then drove back to Eden Prairie for a dump tour before the rally. The ground was muddy, and Lois' high-heeled shoes sunk into it. I was embarrassed. She laughed and said, "The next time you visit me, I'll take you down a muddy path, too." I offered to help her, but she said she was fine and kept going straight ahead at a spirited pace. Symbolic? I think so. This was a woman who let no one and nothing stand in her way when she had a goal in mind.

We returned to the Residence Inn to eat a quiet supper of wild rice soup, bread, and salad with Grant. My mother and the boys joined us for dessert. After we ate, I removed the two gift- wrapped Eden Prairie T-shirts from under the hotel's Christmas tree and gave them to Lois and Will. They seemed appreciative. Time was tight, so we freshened up a bit and drove over to Pax Christi.

During the few minutes ride to church, I expressed my concern that only a few people might come. When we arrived, there were a hundred cars in the parking lot. "You see, Susan, you worried for nothing," said Lois.

As we entered the church, we saw the bus blockade banners hanging from the pillars and heard strains of "This land is your land." Father Tim walked over to welcome Lois and Will and other friends came to meet them, too. I prepared to start the show, but an attack of stage fright gripped me. I quietly confided my anxiety to Lois, who gently looked in my eyes, inconspicuously took my hand in hers, and softly said, "You'll do fine. Don't worry."

She let go of my hand and I climbed the wide stairs to the altar as I had seen Father Tim do every Sunday. I felt a bit like a trespasser in his premises. Finally, I walked over to the microphone to deliver my last speech as an Eden Prairie citizen.

Out of the corner of my eye, I spied Dick Coller jump up and kick the person next to him to do the same. Soon the entire room was on its feet, cheering. It touched me. The soft lighting and warmth of the applause made me feel like a shining star for a few minutes.

I thanked the people for coming, spoke a little bit about Love Canal and Lois, then introduced Will to warm up the audience. "Will is that wonderful voice at the end of the telephone always giving me strategy tips and encouraging me to go for it. His help has been invaluable to our success so far."

A polished speaker, Will used local color in his talk. He said BFI's sins were so bad, they'd curl the ears of Father Tim in the confessional. Will discussed the corporate profile of BFI and how fights like ours are won; not in the courts, but in the arena of public opinion. A combination of humor and facts kept everyone's attention.

Then I introduced Lois. "A gutsy mother who fought to have her family and her neighbors' families relocated from Love Canal before they all died."

Lois took the microphone and boomed out, "You closed the dump! You closed it! I don't think you realize what you've done."

I had heard correctly; Lois Gibbs is a rivetting speaker. For a young woman, who only a few years ago lived on Homemaker Alley baking cookies for her children, she had come a long way. It was amazing to hear such intensity and strength from so diminutive a figure. Certainly this was the woman who forced President Carter to evacuate families from Love Canal.

After she applauded our success, Lois handed me a certificate commending "the Homeward Hills Homeowner's Association for its success in winning a dramatic roll-back of BFI's dumping." She embraced me, and I quickly put the certificate away. I don't think I even said, "Thank you." Never having received an award before, I felt embarrassed and behaved inappropriately. (Later, I apologized.)

Next came "The story."

"I sent my son, Michael, to kindergarten in September of 1978," Lois began, "and by December, he developed epilepsy. I became alarmed as he got worse. Then I remembered reading in the newspaper there was a canal in our area containing hazardous waste . After some investigation, I discovered it ran under my son's school, and there were 22,000 tons of chemicals buried there. I panicked! Then I demanded that my son be allowed to attend another school. My request was denied.

"Eventually, I decided to go door-to-door and see if other neighbors shared my concerns. Very soon I discovered that every household had someone sick. There were cancer, kidney, and gastrointestinal problems, nervous disorders, respiratory problems, miscarriages, birth defects and crib deaths. Children were being born with three ears, double rows of teeth, and extra toes. In our last years at Love Canal, only four babies out of twenty two were normal."

Then Lois gave us a graphic description of her daughter's illness. "Missy had bruises on her body the size of saucers. Her platelets were so low, she would hemorrhage if I put on her seat belt. When she had a bone marrow test, I had to hold my daughter down. I watched as she bled from the mouth."

It was gut-wrenching to hear this horror. Jerri Coller was crying, holding onto her children in the front row.

"Like you," she said, we organized an association and met in churches. We brought our children, too. Then after awhile, there were no more children at our meetings. We had to send them away because they were sick.

"After we realized officials didn't listen to us, we staged protests like your bus blockade. Don't depend on lawyers and scientists. They can't do their job unless you're creating a ruckus. What got us out of Love Canal? Standing together, standing firm, and saying, 'We won't take anymore.' We carried signs outside the gate. We carried coffins to the state capitol. We did everything we possibly could do to muster political clout. And that's the only way we could fight it.

"Your fight has to be similar. You have the opportunity to walk away from this with everything, not like the folks at Love Canal.

"Please, I'm asking you on a long term basis...to get even more involved in fighting for what's important.

"It's important this community wins, from the national level. We need you to win: we in Louisiana, we in Texas, in New York, and Pennsylvania, because we've got to stop this madness all over the country. And if your site gets opened again, if the expansion goes through, it's just going to be that much harder to fight again. It's going to be that much harder on the overall national level to 'plug the toilet.'

"Stand up and fight and give it all you can, because we need to beat this one. It's a precedent-setting case."

A thunderous ovation followed Lois's talk.

After everyone left the church, I closed the building and prepared to leave. Grant was waiting for me outside under the lights. "You really did a wonderful job tonight, Susan. I'm not just saying that; I really mean it." His thoughtful remark touched me.

Several of us returned to the lounge at Residence Inn and talked late into the night. Lois emptied her soul. She said she still worries about her children's health, even though they've moved from Love Canal. They were exposed to so many chemicals, they might exhibit delayed reactions, she anguished.

Lois said she slept little at night because the Love Canal experience wreaked havoc in her life. Lost sleep was etched in dark circles under her big brown eyes. I felt sorry for her. She looked so tired and frail. But despite these difficulties, she worked long hours to help other people in similar situations. She's a most unusual, remarkable, and outstanding woman.

Several years after her Love Canal battle, Lois Gibbs was the first American to win the Goldman Award, an international environmental prize for grassroots leaders set up like the Nobel Peace Prize.

While I shared some poignant moments with Lois, Linda Lehman left the rally headed for home. As she drove a half-hour to her apartment, she thought about her upcoming court appearance. So much responsibility rested on her shoulders. In just a few weeks, she would have to present the backbone of the city's and citizen's case against BFI. If she blew it, they could lose the case. As Linda agonized over her predicament, she spied a pickup truck pull out of a deserted side street and follow her. It flashed its high beams.

"Be cool," she cautioned herself. "You're probably just a little skittish with the court case coming up. Then the truck pulled along side her car. In an effort to lose it, Linda hit the gas pedal and sped ahead. The truck caught up. Next she tried slowing down to ditch the pursuer. This didn't work either. The pickup slowed its speed also.

"No, this isn't my imagination," she thought. Just then she heard shots. They sounded like gunfire. Something hit the rear end of her car on the driver's side. Only minutes from home and very frightened, Linda made a sharp right turn off the highway without signaling, and sped to her apartment. Not prepared to turn, the pickup veered off the road to the left, made a U-turn, followed Linda to her apartment, and parked out front. Linda ran to her apartment and telephoned police who promised to come right over.

Before the police arrived, the pursuer vanished without a trace. The police officer examined Linda's car but found no bullet holes. However, he promised to watch her apartment for any suspicious people hanging around.

"Who would do a thing like that?" wondered Linda as she tried calming herself down. "Would BFI stoop this low to intimidate me?

Next morning, Lois, Will, the Environmental Committee of HHHA, and renegade scientists and lawyers breakfasted at the Flagship Athletic Club. Lois sat at one end of a long table and Will at the other. While we ate, Will encouraged committee members to continue their fight after I left. Later, Will told me he had tried to be convincing. Knowing Will's power of persuasion, I had no doubts he succeeded.

BFI keeps up the pressure

To neutralize bad publicity from the contested landfill case, BFI put out a series of full page advertisements in major newspapers across the country, discussing its exemplary record. BFI's Flying Cloud Landfill was given as an example of how "BFI operates in an environmentally sound manner, correcting acquired problems and preventing future problems."

We were furious when we saw these ads in the Minneapolis *Star Tribune*. Mary Anderson and I dashed off a response, but had difficulty getting it into the paper because it was fairly scathing. We finally convinced the newspaper to give us equal time on the issue, after accepting full page advertisements from BFI containing half truths.

"BFI Can't Buy A Good Reputation," was the title of our response article. We began with: "It never ceases to amaze us what money can buy." Point by point we tore apart the advertisements, then ended with, "BFI can continue to spend millions of dollars on advertisements, public relations, lawyers and experts, but it can never buy the truth. BFI has taken big risks with the people and the environment of Eden Prairie simply to make big money. And that's unconscionable."

Subsequently, people across the country telephoned me about BFI ads in their papers, and I told them they were similarly misleading.

Citizens testify

The contested case was scheduled to begin November 28, 1988, run through December 15, then continue after Christmas until completed. The evenings of December 1 and 8 were set for citizens' testimony. However, people chosen to testify couldn't speak at these sessions, which meant me. It struck me as a good first opportunity for people to take charge while I was still there to encourage or help anyone.

December first, 125 people came to the Hennepin Technical Center for the hearing. Linda Lehman showed up wearing jeans and carrying a box of popcorn. Ric Rosow howled when he saw his scientific expert looking like she was out to the movies with friends.

Ed Crozier of the U.S. Fish and Wildlife Service began the proceedings by articulating why the landfill expansion would adversely impact the refuge. "The vertical expansion will increase the likelihood that Grass Lake will become more contaminated than it may already be." He also called the possible collapse of the bluff over the river valley a "geological likelihood".

Then various citizens came forward to criticize BFI's record. Chris, attorney for BFI, objected on the grounds this wasn't one of the four issues in the hearing. Judge Klein sustained his objection. The crowd began to fume.

"We've been put in a position as citizens that we can't protest the issue of this dump being closed unless we have expertise on four items. I want to register my frustration that we can't talk about protecting the future. The dump should be closed for numerous reasons," said Debbie Belfry, a longtime expansion opponent.

People were intimidated by the Judge's ruling limiting testimony to four issues. Few people volunteered to come forward. Finally, Mary Anderson stood up and said, "You see it everyday and are witnesses. You have more expertise than you think."

Then several engineers in the audience discussed the permeability of BFI's impermeable clay liner. Others finally stepped up to the microphone to comment on the issues also.

"Experimentation with people's lives ought not to be permitted," said one resident who said his credentials included being a father of two children.

In all, 21 spoke out against the dump's expansion. But the highlight of the evening was a four-year-old boy, Greg Swaim, who asked his Dad, our former next door neighbor, if he could speak to the Judge at the meeting. Bill called to ask me what I thought about his son's request. In my experience, children speak from the heart and keep us adults focused on the only issue that counts, their well being. I recommended he let his little boy testify.

At the chosen moment, Bill and his wife, Bonnie, walked up to the podium with their son, Greg, wearing his baseball cap and carrying his teddy bear for support. Bill held Greg up to be sworn in, and the boy promised to tell the whole truth and nothing but the truth. Then he said, "Our family doesn't want to get sick, so please, will you stop the dump? Thank you very much."

At the second open hearing, the public did its homework and came prepared to discuss the issues. Bill Swaim read a letter from an engineer who described a coal mining town in Wyoming where the methane collection system failed. Homes two miles away had to be evacuated because gas entered the basements.

"My hair curled at some of the compounds found in the water," said Marlys Lund, an industrial researcher. "Vinyl chloride, a known human carcinogen, was present in some of the test wells at 2,000 to 2,500 times the EPA-required intervention limit."

Eloquent as always, Senator Don Storm brought the house down with the following remarks: "The premise of our policy should be, first, to do everything in our power to protect the environment of which we are stewards; and second, to do everything in our power to protect the welfare of the citizens. Both are in jeopardy."

I just stood in the back of the room watching both hearings. Mary Anderson acted as spokesperson for the group. I felt like a proud mother bird watching my babies struggle to take their first solo flight. I was pleased to support their fledgling efforts before I left. But as yet, no one had emerged to lead the flock.

Passing the baton

In my heart, I felt the new leader had to be a "stay-at-home" mom. Dedication and time required to accomplish the job, considering all the daytime dump meetings, left no time for a career. Less than two weeks before I packed my suitcases to leave, Mary Anderson, Betsy Nigon, Jerry Coller, and Barb Bohn asked me to sit down and go over everything. They wanted to set up a committee to divide the workload between several people. I explained the case in detail, then turned over my boxes of files to Mary and Scott Anderson, adding my Pennsylvania phone number. Both had demonstrated an avid interest in seeing the crusade to completion. However, Mary was pregnant and worked, and Scott was employed as a sheriff. I wondered how they would commit the necessary time and energy to finish the crusade.

The baton was passed now; it was time to go. The boys and I prepared to leave Eden Prairie, but a round of parties preceded our final adieus. Each boy staged a sleep-over with friends at the Residence Inn. We transformed the loft bedroom into a dormitory, and boys in variously colored sleeping bags were strewn over the floor and on two double beds. They laughed and talked late into the night, happy to be together one last time and doing very little sleeping.

My prayer group at Pax Christi planned a wonderful farewell luncheon. Everyone brought part of the meal in typical Midwestern style, and a prayer was offered for a successful family move. Mary Kennedy also organized a baby sitting co-op dinner with many mothers who participated in our group. We so enjoyed one another as we raised our children together, supporting one another through ear aches, potty training, first days of school blues, and finally the closing of BFI's toxic dump. It was painful to say goodbye to them.

Other friends invited the boys and me to their homes for personal farewells. I remember one evening in particular, when Betsy Nigon, another mother of three sons (birds of a feather flock together), had us over for dinner. Her husband was out of town and Ivan was working in Canada. There we were, two mothers outnumbered by six little boys, aged 3 to 12 years. Betsy and I sat at the ends of her long dining room table, attempting to keep the pandemonium down to a tolerable level, but despite our best efforts, the dinner atmosphere bordered on chaos. We did the only thing we could do in these circumstances: laugh.

192

My departure sparked another variety of party, too: a "good riddance, glad to get Susan Varlamoff out-of-town party" sponsored by BFI and friends. The landfill manager said they planned a big champagne party after I left. Whether he was teasing or not, I couldn't be sure, but I told him not to drink it all; I'd need it for our victory party.

I also said a temporary goodbye to Chris, BFI's trial lawyer. I shook his hand and said "May the best woman win." Then I added, "Since you're not a woman, I guess that means me!" We both laughed. I patted Chris on the arm and told him, I'd be back to testify in court. "You'll miss me, Chris," I assured him, and walked away.

Just days before we left Eden Prairie, Jerri Coller held an open house party at her home for Ivan and me. She put out flyers at the hearings and invited everyone. This would be the saddest party of all. On this occasion, I wanted to give my friends something tangible to remind them of our work together, and inspire them to keep going. I thought and thought, and looked around the shops. Finally, I saw a child's magic wand with a star on the end, all covered in silver glitter. I purchased it.

Ivan flew in from Canada to help us with the move and Grant came over the day of the party. Together, we all went to the Coller's home. Probably 30 people had arrived before us. It was one of those events, I can barely recall, like my wedding day. Jerri had arranged a wonderful appetizer buffet. The "A" list of expansion foes, comprising politicians, scientists, and neighbors, was there. They thanked me for my work and presented me with a signed print of The Cummins Homestead, the pioneer home where I spoke on a soapbox about the dump. I then read them the words of the "Impossible Dream," which I changed slightly to fit our circumstances, while I held the glittered wand with the star:

> *To dream the impossible dream,*
> *to fight the unbeatable foe,*
> *to bear with unbearable sorrow,*
> *to run where the brave dare not go.*
> *This is your quest, to follow that star.*
> *No matter how hopeless, no matter, how far . . .*
> *To be willing to fight,*
> *so that honor and justice may live . .*
> *You must always be true to this glorious quest.*

For your hearts will lie peaceful and strong,
when you're laid to your rest.
For the world will be better for this,
that many men, women, and children strove
with their last ounces of courage
to reach the unreachable star.

Then I handed over the wand to Betsy.

To my great surprise, the Mayor walked in next, carrying a big envelope. He asked for everyone's attention and came over to me. I sat down on the sofa, alone, while he stood over me reading a proclamation from the City of Eden Prairie. "Whereas, the quality of a community is demonstrated by its people, and...."

After the long proclamation was read, the mayor handed me the key to the city and gave me a hug. I was speechless. My mind reeled. Never in my wildest dreams did I imagine receiving any recognition for my work. I just did what I needed to do to defend our children against the potential health effects from a landfill expansion.

Betsy Dick, reporter for the *Eden Prairie Sailor* newspaper asked me what I thought of Eden Prairie. I said that despite the frigid winters, it was the warmest place on earth.

As each of my friends left the party, I walked along to the door. My eyes brimmed with tears as I said goodbye and we embraced. Repeating this parting so many times made my eyes become badly swollen. After everyone left, Ivan and I thanked Dick and Jerri, and Jerri and I sobbed in each other's arms one last time. Parting isn't sweet sorrow, I decided. It's punishment for all my sins of this life and the next.

Ivan packed up our new minivan and we retired our station wagon to the junk yard. Then we headed due east for New York with Neil.

Meanwhile, Pierre, Paul, and I waited with our suitcases and Killy our cat at the Residence Inn for Patricia Madame to take us to the airport. About an hour before she arrived, I received a last phone call from Grant. He wanted to wish me one final goodbye. A BFI lawyer stood beside him, so Grant thrust the phone at him and told him to say a few words to Susan Varlamoff on her last day in Eden Prairie. I sensed this was Grant's idea of a practical joke, because I could hear him howling in the background. However, the familiar BFI voice I knew so well from our days in court wasn't in a playful mood. Instead, to Grant's shock and dismay, the BFI lawyer announced tersely, "If you come back to Eden Prairie, Susan, I'm going to get you."

I was stunned and appalled at this offensive threat. It made me even more upset. When Patricia Madame arrived and picked us up, my dear friend was right when she identified herself as "a damn fool." Who else would volunteer to drive me, in my torment, to the Minneapolis-St. Paul International Airport with one-way tickets to New York?

As I waited in the lounge for the flight to be announced, I wept again. Patricia hugged me and said, "Oh, you'll be back." She was right, of course. I had to testify in six weeks at the contested case hearing. "It won't be the same," I wailed.

I remember nothing after this moment until I reached New York City. Nature must soften pain by allowing the mind to go blank. Ivan met us at the airport with Neil, and I turned over our boys and cat to him. Our cat, traumatized by the airplane trip, soiled her box and smelled terrible. Ivan and the boys washed her in the men's room, then we drove to Grandma and Grandpa's house for Christmas. We turned the final page of our life on the prairie. Or had we?

EDEN PRAIRIE SAILOR (MINNESOTA SUN PUBLICATIONS)

10 | CALM BEFORE THE STORM

I f the *Guinness Book of Records* had a category for the world's biggest cry baby, I'd have earned the distinction for 1989. I was devastated after leaving Eden Prairie; nearly inconsolable. I cried for hours on end. Christmas at my parent's home was barely tolerable. One evening, when Sidney Pauly telephoned, I burst into tears, just hearing her voice.

My condition didn't improve even after we moved into our new home in Pennsylvania. A torrent of tears streamed down my cheeks as I mechanically put up wallpaper and hung pictures on the walls, remembering my house on the prairie. I tried not to cry in front of our children, so as not to upset them, but during the day, when I was alone in the silence of my house without familiar voices of friends calling, I wept.

I must be experiencing Eden Prairie withdrawal, I reasoned. For seven years, I binged on the wholesome life of Minnesota and was now forced to quit cold turkey. The result: the pain of no friends, no phone calls, no fun, and no stature in the community.

Even my usual voracious appetite deserted me. The chocolate-filled

197

air from the nearby Hershey factory nauseated me instead of enticing me to sample its latest confections. As a result, my weight dropped six pounds. Next to the hefty German population inhabiting the area, I stood out like a frail, pale weakling, and I couldn't have cared less.

Then I hated everything about Hummelstown, Pennsylvania. The people seemed unfriendly. My sour disposition didn't inspire anyone to seek my companionship. The Catholic Church services depressed me, I thought the sports teams were badly coached, and felt most of the school personnel not helpful. But, in my twisted frame of mind, nothing and no one could have matched up to what I had left behind in Eden Prairie. (As my mood became more positive, so did my feelings about the community.)

In fact, the town was fairly provincial. It grew up around the Hershey chocolate factory. If you worked for the company, you were in. If you didn't, you were an outsider. Guess where we stood with Ivan working for Boart Hardmetals?

After being at the center of a tornado of activity in Minnesota, this felt like punishment. No one in Pennsylvania knew that my friends and I had shut a landfill in Minnesota, nor could they appreciate the effort it took to achieve this feat. And now after all the energy I had poured into Eden Prairie, I needed to re-establish myself and my children in this new community. It seemed horribly unjust. A sense of isolation and desolation consumed me for months.

Thank goodness for my Eden Prairie friends. They wrote, called, and sent newspaper clippings and reports about the dump activities. How I ached to finish the work I had begun with them, to strive together with these extraordinary people to make a difference in the world. The opportunity seemed snatched away forever. But was it really?

The citizens did telephone, asking my advice, and Grant still valued my opinion in the case. Maybe I could continue working in a support mode, I reasoned. There were probably many things I could do if I just gave them some thought. According to the road map, Washington, D.C. was only two hours away by car. I could lobby the Minnesota senators, representatives, and EPA officials there. I could research health and safety problems occurring at other landfills around the country and send information to the environmental committee. And I could always be available to act as a sounding board for the new leadership's ideas.

It was better than nothing. I wanted so much to see this project through to completion. After 10 family moves, nearly everything I started had to be left unfinished. However, most importantly, I didn't want to be known as a person who stirs up a community on an issue, then skips out of town, leaving friends to mop up the mess.

As Lois Gibbs reminded us, this was a precedent-setting case. Since we were all part of the same human family and could benefit from Eden Prairie's success, it didn't matter from where I gave my help. It only mattered that BFI's dump was closed and cleaned up. A sense of responsibility prompted me to carry on.

Contested case hearing

While I grieved in Pennsylvania over my Eden Prairie loss, a battle of experts took place in a St. Paul, Minnesota, courtroom. BFI assembled the finest group of scientists and engineers money could buy. Our perennial favorite, Dr. Roy ("Baseball"), appeared early in the hearing, dragging along an array of charts and maps depicting his latest portrait of the landfill plume. When Ric and Grant questioned him about the plume's anemic size, Roy said he omitted some data when plotting it, because he felt contamination in several wells came from other sources.

On seeing Roy's map, I noticed there was no plume to the north and west of the landfill near the drive-in theater. Then I remembered the time Dick stood up at an MPCA meeting and insisted the theater was polluting the dump. Dick always provided us with comic relief at these otherwise boring affairs.

Many more BFI witnesses with assorted letters after their names followed Roy, each claiming the dump expansion was engineered well and was safe for nearby residents.

MPCA's staff followed with a similar line of reasoning. Scott Fox, staff hydrogeologist, said contamination wouldn't enlarge if more garbage were piled onto existing unlined portions of the landfill. (Maybe Louis was right when he once angrily declared, "The MPCA does nothing more than license pollution.")

Linda Lehman appeared next on the witness stand. Unfortunately, the stress she experienced anticipating her court appearance put her in the hospital several days. As our chief scientist, she knew Chris would cross-examine her for hours on end, hoping to wear her down.

199

Justifiably, Linda was anxious and so were we. Jerri had neighbors send her get well cards, and I called Linda the night before her appearance to encourage her. Patricia Madame even came to court to give Linda a big hug before she took the stand. Outfitted in a red dress, Linda looked like Little Red Riding Hood ready to meet the Big Bad Wolf.

Low keyed and quiet at first, Linda staunchly defended her graphs, charts, and models for two solid days. Her version of the plume showed water contamination double the size of Roy's model. It extended beyond the entire perimeter of the dump, ran under the adjacent neighborhood and drive-in property, and down to the river. Linda explained she used all the data, even the latest sampling results. By the time I reached her the evening of the second day, she was finishing off a bottle of champagne and laughing, very happy this ordeal was over.

Matt Walton, retired director of the Minnesota Geological Survey, succeeded Linda. He displayed the United States Geological Survey map of Minnesota to demonstrate that BFI's landfill sits in a dry depression, which has been unable to hold water since the last glacial ice age, 10 million years ago. "That means this area had extremely porous, permeable materials underlying it, allowing water to penetrate and disappear."

"This site does not have the ability to keep waste isolated from groundwater," Walton said.

"Nothing I've seen in the plans for developing this expansion gets at the fundamental problem, which is that BFI's landfill has a big leak in the middle of it."

After all our city's experts concurred that expanding a toxic, leaking landfill was an ignorant idea, we, the landfill's neighbors, got our day in court. Our association's attorney, Grant Merritt telephoned often to keep me abreast of citizen testimony, and said everyone was holding up remarkably well.

Bill Schmidt relished the opportunity to tattle on his former employer. He described the 1987 washout along the bluff that left garbage exposed, and explained how he covered it over with dirt as he was asked to do. This lead us to believe there was still trash buried at the edge of the slope. Bill also discussed how, on another occasion, he dumped garbage within 100 feet of the bluff's crest, beyond the site's boundaries.

200

By mid-February, it was my turn to testify on the issues. My dear Dad, the man who taught me to love and respect nature, agreed to care for his grandsons while I flew off to defend it.

It snowed February 13, 1989, the day of my court appearance. Sever Peterson, scheduled to speak before me, offered to drive us both downtown in his four-wheel drive vehicle. I felt honored to ride with Sever, a young blonde farmer of Scandinavian heritage whose family plowed the fertile Minnesota River Valley for three generations. He had a B.S. in Agricultural Economics, served one year in Vietnam with the International Medical Relief Agency and sat on the Board of the Hennepin County Soil and Water Conservation District. As we drove through the blizzard, we talked about why we both supported the anti-expansion crusade. To protect the necessities of life, our families, and our land, we decided.

It gave me a warm feeling on a cold day to share experiences with Sever. Crusades like ours must have drawn forth the creme of society, I thought, because I met so many high-minded people fighting BFI's landfill expansion. My admiration for them could fill volumes, and now that I had left Minnesota, their goodness was more evident. I missed them so much.

When we arrived at the MPCA, we parked the car and walked briskly, heads down into the wind, to the entrance doors. Grant, Ric, Joe, the Mayor, and several HHHA members were there to greet and support me. It made my day. Chris, BFI's trial lawyer, waited for me too, ready to devour me.

I smiled and said, "Hi Chris, how are you?" He didn't smile, but nodded. I guess the hungry lion didn't want to get too cozy with his victim.

As I glanced around the courtroom, I saw a weary lot of drawn and haggard lawyers. For two months, these men had been trying the case by day, and spending long hours every night preparing for the following day's witnesses. Judge Klein appeared to suffer the worst. He kept a pillow behind his back to support his spine while he sat for hours everyday listening to tedious scientific facts. When he stood up during the recess, he walked slowly, like an arthritic old man, and not the 40-year-old his face suggested.

Sever was scheduled to begin the day's testimony. After he was

201

sworn in, he drew the landfill bluff on a blackboard, and sketched in five landslides that had occurred over the years. Then he discussed the condition of the watercress at the foot of the landfill bluff. Sever said for many years, members of his family gathered it from the little streams to eat, but not anymore. "Three years ago there was a turbidity in the water, a chalk or clay. We stopped cutting watercress at this time." Interesting, I thought. Since a farmer knows the land intimately, he doesn't have to test for contamination. Changes he sees in vegetation indicate something is wrong.

After Sever concluded his watercress story, I walked over to Judge Klein, raised my hand and promised to tell the truth also. After I sat down, BFI's lawyer, Chris, began addressing me as Mrs. "Varmaloff." Grant quickly cut in and objected. He insisted Chris get my name right. With as many times as I appeared before him, Grant said, he should be able to pronounce "Varlamoff" correctly. I laughed, relaxing a little.

Sitting on the witness stand gave me a closer look at Chris. He reminded me of a spider, a "daddy long legs," because he was tall and thin and had unusually long limbs. From time to time, he extended his arms and legs in all directions to stretch, but he wasn't doing that now. His body seemed tense and tightly held together when he continued after Grant's interruption.

Chris now queried me about my educational background and work experience, then asked me to testify on the four issues. After checking my prepared notes, I answered, "Yes!" on the first issue, whether expansion will significantly exacerbate existing contamination. I qualified my response by explaining we don't really know how bad contamination is under the dump, because BFI tests water only around the periphery. Then BFI's scientists extrapolate to determine the levels in the center. Chris objected to my challenge of his expert's testimony. He said, I didn't have the credentials to make this judgement.

I objected to his objection of my testimony and meekly asked the judge if I could defend myself. He gave me permission to explain my reasons, so I discussed my work as a research assistant at Lamont Doherty Geological Observatory testing deep sea sediment samples for quartz and opal. We always ran several analyses on many samples evenly distributed over an area of the ocean floor, I pointed out. Only when Chief Scientist Dr. Biscaye was satisfied we had sufficient data did he draw contours identifying the varying quantities of minerals on the ocean floor. This is standard scientific procedure, I explained.

Regarding the second issue, whether the landfill's liner and leachate detection system specifications are adequate to prevent leachate from migrating to groundwater, I said, "No!" and recited the second law of thermodynamics (A scientist advised me on the importance of this argument): "The entropy of a closed system can never decrease. In simpler terms, nature tends toward disorder, which means the liner will eventually break apart and leak," I confidently commented.

On the last issue, I objected to the wording. "It's not whether the expansion of the landfill will significantly increase the potential for failure of the bluff, because the bluff will fail on its own anyway. It just depends on how much garbage we want in the river, 18 years' worth if the landfill stays shut, or 50 years' worth if the expansion is permitted."

I spent an hour and a half on the stand and gladly yielded my place to Alex Zubar. After he finished, we all went out to lunch. Alex and I ordered wine to toast our survival of the trial. I returned to court giddy, and found Chris exceedingly funny that afternoon. Before long, I grew tired and rested my head on the table, nearly falling asleep. Later, I wondered what the Judge must have thought of my sluggish behavior.

During my stay in Eden Prairie, I resumed my 14 hours a day work schedule, fighting the expansion as though I had never left. One day, Sidney invited me over to the legislature to hear a presentation by William Rathje, an archaeologist who exhumed landfills. How could I resist an opportunity to learn more about my favorite subject: dumps? I gladly accepted Sidney's offer. As we sped around state capitol halls, she remarked to prominent political passers-by: "Susan is still fighting the Flying Cloud Landfill expansion." (I imagine Sidney wanted to impress the opposition that even though I was out of state, the crusade wasn't out of my mind.)

We arrived early for the lecture, so I seized the opportunity to ask the speaker what he thought about expanding the low-tech, leaking dump in Eden Prairie. "Those old landfills all leak and need to be closed," he said. When he spoke for the record, he brought up his conversation with me and recommended that the 1970's era dumps be shut down. "Another nail in the coffin," quipped Sidney as we merrily left the room, pleased with ourselves.

I also met with the Environmental Committee to help them any way I could. We had a good turnout for the meeting at Scott and Mary's home, but the group appeared to be floundering a little. I could understand why; it took me years to develop the confidence and skills to lead people.

Barb Bohn chaired the meeting, and Scott briefed us on the latest in the contested case. He talked daily with Grant and Linda, and gathered piles of technical information he studied and presented to the group.

At this meeting, I shared with them a psychological ploy I used in dealing with the landfill issue. "Never, under any circumstances, show any signs of giving in or giving up to the opposition. Act confidently you'll win, and tell BFI you'll stop this expansion with whatever it takes, for as long as it takes.

"If you repeat this tactic, you'll convince yourself and others fighting alongside you that victory is inevitable. And it'll shake BFI into believing they'll lose." Several years later I read the words of Winston Churchill who said nearly the same thing during World War II: "Never give up! Never! Never! Never! Never!."

Eventually Jerri and I laughed together as we reminisced about those times when the two of us were so exhausted from fighting this crusade that our complexions turned chalk white, and dark circles seemed permanently etched under our eyes. How tempted we had been to slide down a wall and crumple into a heap of defeated humanity on the floor. Instead, for the sake of appearances, we gathered our flagging energy and strutted around meetings holding our heads high and crisply saying: "We're going to win; it won't be long now." We did this for three years.

Before I left Eden Prairie, I took a nostalgic trip to Target, where I had spent thousands of dollars over the years saving money on their sales. As I strode down memory aisle, a woman I never met before stopped me. She had tears in her eyes. I stood confused. Then she profusely thanked me for all the work I had done for the community, and began enumerating my deeds one by one. I felt ill at ease, but listened and expressed my appreciation for her remarks. After we parted company, a feeling of satisfaction filled me, and I felt better when I returned home to Pennsylvania.

By the end of February, 1989, the contested case was over, which meant Judge Klein must sit down, reread his notes, and make recommendations on each issue. He declined to take a position for or against the expansion; that was the board's decision, he said.

Our lawyers worried about the outcome of the case. Of the dozens of issues we requested to be included in the hearing, the MPCA chose only four. These few were worded in a manner that made it very difficult for us to win. However, MPCA regulations did list "fitness of the applicant" as a criteria for obtaining a permit. The staff told us the Board would consider Woodlake's record independently of the contested case issues before they made a decision on the landfill's expansion. Both Judge and lawyers discussed the fitness question and concluded that legal procedure required the expansion decision to be based solely on the hearing issues. Panic set in, but a light bulb flashed on.

We all knew BFI's record nationally was horrendous, and exposing it could give us the edge we needed to clinch the case. Our lawyers headed over to the MPCA to speak with Eldon Kaul, an assistant attorney general for the state of Minnesota, who was absent when the staff recommended the first four issues. The staff had explained that compliance could no longer be considered under the contested case rules, but the MPCA normally examines the operating record of an applicant before issuing a permit. Kaul reported this discrepancy to the board. Board members said they misunderstood the procedure and wanted BFI's past operations record examined. So lawyers from both sides filed briefs outlining their positions.

Ric Rosow prepared a 25-page memorandum for the judge, explaining why BFI's corporate record should be exposed. Up until now, landfill officials insisted that Woodlake Sanitary Service, Inc. operated as an entity apart from its parent company, BFI, the culprit we so often associated with so many national crimes. "Not so," wrote Ric. "The basic contention is that WSSI and its BFI parent are so intertwined as to be indistinguishable." He cited examples of how John Curry, a Woodlake manager, referred to himself in his deposition as a "BFI" employee, and Gerald Lynam, President of Woodlake, explained that funds collected from the subsidiary are placed into a depository account from which Woodlake has no authority to withdraw. Ric commented, "If BFI and Woodlake are not connected, then why does the sign at the landfill entrance say 'BFI,' and why do all the blue trucks coming out of the dump have 'BFI' painted on their sides?"

Ric wrapped up his report with 17 pages of violations, fines, grand jury investigations, and court cases pending against BFI. I imagine this project kept him awake a few nights past midnight.

Chris, BFI's trial lawyer, filed an eight-page document opposing introduction of BFI's record into the case, saying, "Courts are typically very hesitant to 'pierce the corporate veil,' and will do so only in extremely compelling circumstances."

Compelled to know the truth, the PCA Board voted seven to one to add permittee suitability to the hearing.

This news must have rocked the corporate foundation of BFI headquarters in Houston, Texas, like an earthquake. Shortly after the ruling, Grant called me, very excited and announced: "Ruckelshaus is in town. He's meeting with all the government heads on the Flying Cloud Landfill expansion."

Even though we were told our case topped the company's priority list, we hardly expected the CEO to fly into town. Grant frantically tried tracking him down for a meeting, but his old pal, Bill, was too busy to return calls. On his grand tour, Ruckelshaus contacted the Eden Prairie City Council for a meeting, but they became occupied with other matters and declined his invitation.

Funeral procession

Pleased with their ability to enlarge the contested case's scope to include applicant fitness, anti-expansion forces decided to go for the gold, the addition of three more issues:

1. Whether the landfill was suitable for the proposed expansion.

2. Whether airborne emissions of methane and other landfill gases from the expansion would adversely affect the environment.

3. Whether venting of contaminants from the site's air stripper would have an adverse effect on the environment or constitute a potential health hazard for residents of Eden Prairie.

Linda Lehman prepared a report to support the need to explore these concerns and kindly sent it to me. This was probably her best work; the cornerstone of our case against BFI and the government.

In it, Linda reviewed calculations the MPCA made to arrive at a Superfund ranking of 40 for the Flying Cloud site. She found some serious omissions; the staff didn't assess the air pathway portion of the

206

score, which determines potential harm to humans, and they didn't ascertain the fire and explosion score. Both these factors posed the gravest threats to health and safety of BFI's landfill neighbors. Linda re-evaluated the site using the U.S. EPA Hazard Ranking System, and discovered Flying Cloud Landfill ranked as one of the most hazardous sites on the Minnesota Superfund List, and deserved a place on the National Priority List.

"Populations within a four-mile radius are an indicator of those who may be harmed should hazardous substances be released into the air," Linda wrote. "There are over 22,000 people living within four miles of this landfill, and they're 'targets' at risk to exposure. Negligence of this pathway is inexcusable, due to the explosive and incompatible nature of chemicals contained in this landfill."

"Ironically," she noted, "the MPCA has scored this facility as representing a significant risk or hazard. Yet, it now indicates there's little or no risk to the expansion."

Linda went on to say the site violated solid waste management rules and policies of the U.S. EPA and the MPCA, and, therefore, was clearly unsuitable for an expansion. Par for the course, I thought, as I read her words. In 1987, the Metropolitan Council also ignored its own policies and pushed through the expansion.

As I read Linda's report, it became clear that recent problems with sanitary landfills across the country resulted in a flurry of new government legislation we could finally use to support our position. Back in 1982, when we began our anti-expansion campaign, there existed few laws and little scientific data on which to base our opposition. No wonder our fight was taking so long. Without dead bodies in the street, we had a less dramatic basis of fact to support our case. At the time, expanding a leaking, toxic landfill just seemed like a dumb and dangerous proposition. But as the French say, "Time tests the fabric." When BFI's landfill leaked, it confirmed our assumption that the site was worthless.

Originally, when we petitioned the staff to include site suitability as an issue in the contested case, government officials refused to enter it, because "a permitted landfill already exists at this site." They said Flying Cloud wasn't considered a new site, just an expansion of an existing landfill. Therefore, it didn't have to adhere to current regulations for

207

suitability.

The MPCA insisted contamination wouldn't increase at the dump, because BFI intended to remediate the current groundwater pollution and prevent off-site methane migration with engineered solutions. The EPA said, "They (engineered measures) cannot be relied upon to make an obviously vulnerable site, such as a sand and gravel aquifer, non-vulnerable."

"It is simply wrong to depend upon an engineered barrier well system, which the operator is only required to maintain for 20 to 30 years, to protect the population and environment from the landfill emissions for hundreds or thousands of years," concluded Linda Lehman, Eden Prairie's Scientific Consultant.

As yet, there was no groundwater remediation system at the dump, even after BFI officials signed a consent order in 1985 to install it. Second, we obviously had no guarantees it would work properly, since it wasn't operational. And transferring chemicals from water to air, using an air stripper, defied our comprehension as a clean-up solution. If these carcinogens represented a hazard in water, how could they be harmless in the air? Lastly, the barrier well clean-up system would be installed on the crest of the bluff, the same bluff that was periodically eroding into the river. It didn't take a Ph.D. in anything to figure out the final destination of the groundwater remediation system in time.

In addition, there were still concerns about how methane and other gases would impact health of nearby residents.

In April, 1988, after BFI's landfill shut down, a BFI official was quoted in the *Minneapolis Star Tribune* as saying: "The company said it also will spend at least $2.5 million to convert the gas it collects into electricity. Our plan is to have it in place as soon as possible. We think this will put Flying Cloud on the forefront of landfill operations."

Unfortunately, they never got around to it. BFI officials told Judge Klein, after the MPCA staff decided the contested case issues, there was no current plan to convert methane gas produced at this landfill into either electricity or heat for homes or buildings. (It appeared the government and BFI shared the same basic philosophy: promises and rules are meant to be broken.) The company decided to flare the gases instead, but how could we be sure they'd all burn and there weren't toxic gases in the flares?

The United States Office of Technology Assessment issued a special report, *Are We Cleaning Up?* in which it questioned the whole idea of landfill remediation. "Are we cleaning up the mess or messing up the clean-up? Superfund clean-ups remain largely ineffective and inefficient," it concluded,

After reading this and many other reports, I sent Grant several pages of arguments to use for requesting additional issues, but the stupidity of the exercise disgusted me. I closed my letter saying, "This whole process is a plain embarrassment."

Chris petitioned against adding the issues and wrote, "The applicant opposes this delaying tactic . . . " (Good grief! Could BFI be catching on to our strategy?)

Our lawyers prepared excellent briefs, quoting a meeting transcript where several members of the MPCA Board requested information about site "topography, geology, hydrology, and soils." We counted on them to follow through with their demands. However, we never left fate to scientific arguments or the good sense of government officials, so we planned another citizen demonstration just before the decision on the additional issues.

Scott Anderson and I brainstormed for some new idea, but came up empty. So I called Will and he asked Lois. She suggested a funeral procession to the state capitol. The group loved it. But I hated my phone bill, which topped $100 every month now.

The Environmental Committee plunged into preparations. Mark Utne, a new member, made two pine coffins; a small one for a child and a full size model for an adult. Two neighbors who owned black vans volunteered to transport the coffins downtown and several children and adults came forward to act as pallbearers. The public relations moms got out flycrs and notified the press. Everything was set for April 16, Earth Day, except there were no speakers.

The committee asked me if I'd address the people. Of course, I wanted to stay involved, but I thought they should find someone else from among our group to speak, too. I agreed to return on condition they come up with their own spokesperson. After much debate, Cary Cooper reluctantly accepted the job.

I wanted to write a blockbuster speech, so I went to the library and xeroxed the *Gettysburg Address* and Martin Luther King's, *I Have a*

Dream speech to use as models in preparing my own address. After intense writing and rewriting, I boarded a plane to Minneapolis once again, hand-carrying my speech, while my father babysat again for my boys, and Brenda Gowan tolerated me one more time as her house guest. I arrived several days before the demonstration so I could help with other dump matters.

Dwain Warner and I managed to schedule a meeting together during my stay. I asked him to give me a guided tour of the river valley. He happily obliged, and Jeff Gowan accompanied us. Once again, the day was deliciously beautiful. What fun it was to learn from the expert himself about grasses and animals inhabiting the area.

When we concluded our tour, we adjourned to the local pie shop and discussed our case. I enjoyed Dwain. His intelligence, cunning, and great sense of humor always entertained me. As we spoke together, Dwain drifted off, his eyes darting around the room. He whispered, "You see that man over there? I think he's a spy. Keep your voice down."

Now that's a challenge for someone who likes to talk loudly and at length. Since Dwain seemed serious, I obeyed his order. Then he quietly said he had aerial infrared photos of the bluff, taken twenty years ago and today. He felt sure we'd see evidence of dying vegetation on recent slope photos, due to methane gas seeping through the soil and killing the grasses. This methane migration would dramatically increase the failure of the bluff, he said, supporting our theory that the dump would increase instability of the slope. It was an excellent idea. I told him to pursue it. He asked only that his expenses be reimbursed. I proposed to speak with the citizens about it; I couldn't see them objecting.

I also asked Governor Perpich's aide for a meeting between the citizens and the Governor. Mary Anderson and others wrote letters requesting this opportunity too. Mary remarked to the newspapers that when BFI's CEO Bill Ruckelshaus flew up to Minnesota, he was given immediate access to Perpich, while the citizens, who live by BFI's landfill were made to wait years for an audience with the Governor.

The grand total of our collective efforts produced an appointment with Rudy Perpich several days after my departure. This arrangement deprived me of a chance to meet the Governor, but I knew our committee members would present their position well. I went over key technical matters with Cary and Scott before I left, so they'd be well prepared.

Since they represented current leadership, it was better they spoke for our group. However, I had strong suspicions the citizen get-together was intentionally scheduled after my departure. I later learned the government viewed me as a rabble-rouser, and may possibly have wanted to avoid me.

Sunday, April 16, was a splendid day, so I walked up Homeward Hills Road to Pax Christi Church for Mass, and remarked that the sun shone on every one of our outdoor protests. Was this a good omen? I felt certain it was. Like the words of the Girl Scout song I sang too many years ago, "All night, all day, angels watching over me, my Lord ...."

When I walked out of church, I saw Jerri Coller and her husband, Dick, and went over to greet them. We talked about the afternoon rally, and Jerri said she had worked hard drumming up support for it in her Deerfield subdivision. I had complete confidence in her ability to get her neighbors downtown.

After I returned to Brenda's home, I noticed the press across the street at the Anderson's home. I watched a pregnant Mary do an outstanding job speaking to reporters. It saddened me to stand on the sidelines, but, on the other hand, it also made me proud to see how well Mary handled herself.

After lunch with Brenda and her children, we all rode up to Pax Christi Church to join the others for the funeral procession. Meanwhile, as the Collers prepared to leave for the rally, they realized all their neighbors, after promising to participate, had no intentions of doing so. This being the first warm day of spring, everyone decided to scrap the protest and invite friends over for the year's first backyard barbecue. Feeling betrayed, Jerri stormed into neighbors' backyards making comments like: "Sure hope you have fun at your barbecue," or, "I hope you choke on your hot dogs!" Then she returned home and told Dick what happened.

Dick took matters into his own hands. He recircled the neighborhood telling neighbors they let his wife down. "The dump will stay open with your attitude," he loudly warned. With that, the Coller family departed in their black van with their two children and a coffin, to join the others at Pax Christi Church.

I stood around the parking lot with Jerri, nervously waiting for more people to arrive; there were very few cars until now. Then, as if by magic, a stream of minivans and cars drove up Homeward Hills Road and pulled into the church lot. Jerri recognized the group: the Deerfield barbecuers! Feeling guilty, they shut off their grills, put away their hot dogs, and invited their friends downtown to the rally instead of a backyard picnic, which could come later.

Now 75 cars revved their engines, put on their headlights and formed a line behind the Coller's black car. I rode in Brenda's station wagon somewhere in the middle of the pack. We snaked slowly down the right hand lane of the beltway, to the Minnesota State Capitol. Many processional cars carried signs in the windows explaining our mission to passers-by: "PCA, PROTECT OUR NEIGHBORHOOD!" and, "BFI, DON'T DUMP ON US!"

When our motorized parade arrived at the capitol, cars parked, coffins were unloaded, and people organized themselves for the march. I walked ahead, alone, up the long flight of stairs to the landing where Senator Don Storm waited to begin the ceremony. Just once, I wanted to watch a demonstration.

As I stood under a 75-degree sun, I realized I miscalculated in choosing clothes for the day. A green turtleneck worn under a long, black wool sweater complimented my black dump skirt but roasted me. I suffered for the sake of being color coordinated for this earth day funeral procession.

Starting with pallbearers, protesters lined up behind the Greenpeace banner, "BFI OUT," we used in our bus blockade. Children outfitted in school sweatshirts picked up the small coffin, and men and women dressed in environmental sweatshirts carried the large coffin. Slowly they marched up the central promenade.

It was impressive to see the line of people quietly make their way to the capitol steps. Dick coached the young boys struggling under the weight of the pine box. I recognized the children; several were my son's friends. Choking back tears, I remembered the time I saw another small coffin pass before me at my sister's funeral.

The group stopped in unison when Don gave a signal; the caskets were lowered to the ground. Then, Senator Storm, formerly a minister, began the ceremony with a prayer, and we bowed our heads in respect. Jerri and I put our arms around each other's waists during this moment, and Jerri cried.

212

After the prayer, Don introduced the speakers: "We will hear first from Sidney Pauly who is Eden Prairie's well loved House representative." While she spoke, Cary Cooper came over to me; he was getting nervous. I held his hand like Lois Gibbs had done for me while we waited to speak. After Sidney's speech detailing all the problems at the landfill, Don called me forward. "Next, we'll hear from Susan Varlamoff, who has been on the cutting edge of this environmental fight."

As I began speaking, I noticed Carol Culp standing at the back of the crowd, smiling at me. Several years ago, she attended our Ring Around The Dump with her family, and now was here again. It was reassuring to see her friendly face and many others beaming at me as I boomed out: "Like the rag tag army of American patriots who slew the mighty British forces with not much more than a desire for freedom and justice, we, a poor army of citizens have taken on the multi-billion dollar corporate giant, Browning Ferris Industries. Their legions of lobbyists, lawyers, and experts seem no match for our bedraggled army, but we, too, have the burning desire for life and environmental justice. And we are armed with the most powerful weapon of all, the truth . . . Let there be no misunderstanding, we will stop at nothing to ensure that this landfill will never reopen, because we know leaking dumps kill."

However, the best speaker was last. Don introduced Cary as someone who spent 25% of his life attending more than 100 meetings in the last seven and a half years to block the expansion. Cary focused all his nervous energy into his speech resulting in a command performance. First, he threw out questions to the audience to get their participation.

"How many of you are frustrated?"

"Everyone!" shouted the citizens.

"Will we allow an expansion of a dump at a terrible sight by a terrible corporation?"

"No!"

Then he continued: "We have been Hennepin County's garbage can for too long. The can is shut and it will not be reopened."

Wild applause!

"We will continue to demand environmental justice. We will continue to set the record straight. But we will not stop there. It is our desire to make this funeral an Irish wake. It is time to set up the dead and to celebrate the living. We will invest countless additional hours to re-establish trust in government, and live in our community without fear. We will continue to fight for ours and everyone's environmental rights."

That put an end to my concerns about our citizen's inability to speak for themselves.

We concluded the ceremony by singing, *This Land Is Your Land.* Afterwards, the demonstrators drove back to their homes, and I accompanied the Collers to an ice cream parlor, followed by a visit to Minnehaha Falls for a delightful few hours before taking a plane back home.

Meeting the Governor

At the appointed time and day, Scott Anderson, Cary Cooper, Betsy Nigon, Jerri Coller, and Grant Merritt walked into the Governor's office with their list of demands. "The meeting was relaxed," Betsy wrote; "he (Governor) seemed sincere and interested, although he does not impress me as a political official."

Cary and Scott gave the history and technical aspects of BFI's toxic, leaking landfill, and expressed frustrations over the Minnesota Pollution Control Agency staff's recommendation to expand the landfill. "We indicated to the Governor that if there were a concern, (the MPCA) should jump on the bandwagon of at least going overboard to protect its citizens, rather than BFI," Cary said.

At that moment, our citizens were also worried about a bill pending in the legislature that could reduce the inventory of proposed landfill sites, putting pressure on the government to expand the Flying Cloud dump. This was the same issue we defeated the year before. Our citizens asked the Governor to look into the bill and let the legislators know he wasn't interested in removing other potential sites. Perpich seemed to support this proposal, Cary indicated.

After our representatives finished their presentation to the Governor, they asked him to visit the site. He agreed. They felt if he could see the landfill's proximity to the neighborhood, airport, and wildlife refuge, he'd realize this wasn't a good spot for garbage.

At some point, Grant presented a letter from me in which I requested the Governor to support Eden Prairie citizens on this potentially grave matter. Grant later told me that when he presented my letter, he told

Rudy, "You have just lost two outstanding citizens of this state, Susan and Ivan Varlamoff."

Our representatives were satisfied with their visit. No longer could the Governor claim ignorance as an excuse for inaction.

MPCA and legislative decisions

I impatiently waited all day April 25, 1989, for a telephone call from Minnesota to tell me the MPCA Board's decision on the contested case additional issues. If these issues were granted, we knew we had a good chance of winning. The day dragged on as I imagined how the meeting would go.

Some things were predictable: school buses would take mothers, children, and their paraphernalia downtown, and the men would come running into the meeting from their offices to make an appearance. Chris would stand up and say, "There's no need for additional issues, all the citizens' concerns were addressed in the first four." Then, Dwight Wagenius, MPCA attorney, would agree. Lastly, Ric, Grant, Linda, and Barb Bohn, spokesperson for HHHA, would charge forward denying that all their concerns had been answered, and present information supporting additional issues.

My day passed as usual. The boys went to school, returned, had snacks, watched TV, ate dinner, did homework, went to bed. I was the only one who didn't sleep. (Ivan was out of town.) I read with the phone at my elbow; ready for a quick response.

It rang about 10:00 p.m. and I snatched it up.

"We won!" exclaimed Grant.

"That's fantastic," I cried. "Tell me everything."

As Grant related events of the day and how well Barb Bohn spoke on behalf of the people, a beep sounded in the phone. I put Grant on hold and switched to the other line. It was Barb. She was ecstatic about the decision and wanted to tell me.

As it turned out, the board approved *four* new issues; one more than we had asked. Apparently, the board wanted more information on how the expansion would impact the refuge and added it. The MPCA Board, comprised of individuals with varied backgrounds, including law, medicine, biology, environmental activism, and farming, was evidently disenchanted with the staff and decided to take matters into their own

215

hands. Ironically, the only dissenting vote on these mostly health and safety issues was the doctor who chaired the meeting.

Occurring simultaneously with the contested case hearing was the legislative battle. There were two bills before the house and senate involving BFI's Flying Cloud Landfill expansion. The first contained a provision permitting garbage to be transported from one county to another for landfilling. Passing it would take the pressure off expanding the Flying Cloud site. We lost on that bill. The other eliminated the last three proposed landfill sites.

Sidney continued passing out information in the House exposing BFI and the landfill site for what they were: bad news. Don covered the Senate, and Grant, along with the citizens, lobbied day and night as we'd done the year before. Again, we lost in the Senate, but by a lesser margin, and won in the House. Again, the bill went to a conference committee for resolution. At 2:36 a.m., Cary Cooper and Grant Merritt walked down the state capitol steps as victors. I spoke with Cary soon after. Yes, he said, it was a mountain top experience.

Taking the case to Washington

Hummelstown, Pennsylvania, is approximately two and a half hours by car from Washington D.C. Wonderful, I thought; I'll drop into town and visit Minnesota's U.S. senators. Not so easy, I found out. I called to make an appointment with Senators Durenberger and Boschwitz, but was told this wasn't possible; first, I must speak with their aides. It was better than nothing, so I headed out the door one day in March at 7:00 a.m. for my first appointment at 11:00 a.m.. (I always give myself ample time to get lost.) Dressed in my six-year-old red Easter suit, I carried my map, dump papers, and bag lunch on a trip to the capitol.

I hadn't traveled to Washington in a good 20 years, and, I reiterate, driving isn't my strength, especially in heavy traffic in unfamiliar territory. With tense hands, I turned on the ignition, gripped the steering wheel, and began the trip. My plan was to park at the Metro Station and take a train into the city. First I had to find the station.

Unfolding the map next to me with the route highlighted for quick reference, I steadily pressed forward, going south on I-83, singing golden oldies with the radio. As I approached Washington, the highway became congested and, as luck would have it, I got lost. It was close to 10:00 and

216

I hadn't yet found Metro Station. Quickly, I asked directions at a fast food restaurant. Dashing out of my car and frantically waving an opened map, I asked a sympathetic looking man the route to the Metro. "I'll never find my way!" I gasped.

"Yes, you will," said the older gentleman, very reassuringly. "Lobbyists always find their way to the right office in Washington."

He gave me good directions and I was off and driving. I soon found the station and parked. Next, I had to decipher the inner workings of the Metro system. Looking forlorn, I asked a young man for instructions on how to catch the train. Afterwards, I discovered he was a Georgetown law student. He showed me how to buy a ticket and pass through the gate. I followed his example and sat near him on the train. Referring to the Metro map mounted on the wall inside the train, he showed me where to get off.

I exited at the correct stop and raced down the street to an immense white building where Rudy Boschwitz had his office. I was running late. Security was tight, so I had to pass through a metal detector before I could go inside. Then I had to find the senator's wing and hallway. Arriving with no time to spare for my 11:00 o'clock appointment, I breathlessly gave my name to the secretary, and was told to wait for Joseph Pendal, Senator Boschwitz's aide. This gave me an opportunity to sink my body into a nearby leather chair.

The surroundings were opulent. Fine wood paneling covered the high walls, and the heavy upholstered furniture was top quality. Pricey artwork and photos of the senator with various high government officials adorned the walls, and the secretaries were young and beautiful. On my way in, I noticed similar offices where other government servants toiled for their constituents, or so they said. However, something didn't feel right about this picture; serving the common folk in such luxurious environs seemed hypocritical. Shortly after my visit, scandal after scandal was revealed on "the hill," involving impropriety of personal funds. As a result, the American people got angry and began a grassroots campaign to "throw the bums out."

Telling the landfill story to Mr. Pendal was uneventful; so much so, details escape me. He didn't seem particularly concerned, but listened politely and took some xeroxed information I supplied. I received no indication Senator Boschwitz would do anything for us. It wasn't long before I found out why. My big-moneyed competition Irv, BFI's Washington lobbyist, ran a big election fundraiser for Senator Boschwitz. However, it made little difference: the senator never got re-elected. He became a casualty of the "throw the bums out" campaign.

After my meeting with Joseph Pendal, I left the building and stood outside on the great marble steps, eating my peanut butter and jelly sandwich and drinking my boxed apple juice. I suspect my behavior was totally gauche in this sophisticated town of well-dressed lobbyists and international diplomats, but I didn't have extra cash to spend or time to waste looking for a restaurant, so I brown-bagged it.

After my ten-minute lunch break, I looked at my street map and found the Russell Senate Office Building where Senator Durenberger conducted his business. It was only several blocks away, so I decided to hoof it. Once there, I passed again through security and proceeded to take another long walk down another long hallway to the Senator's door. If nothing else, I got my aerobic exercise for the day. Inside, a young blonde secretary, no doubt a Scandinavian import from Minneapolis, pleasantly asked to help me. I stated my business and sat down in another comfortable, poshy chair in another sumptuous room, awaiting a staff person named Ed Garvey.

When Ed came out, I enthusiastically re-explained the landfill case with my multitude of papers. The story amazed him. He was equally impressed I had traveled from Pennsylvania for the day to discuss the situation with him. I had hoped that meant he'd take me and our problem seriously.

Unfortunately, for the time being, Durenberger was embroiled in a big headline-making scandal that brought him before the Senate Ethics Committee. How much he could do for us, I didn't know. These days, his staff concerned themselves chiefly with controlling the fallout from the negative press that made the front page of the *New York Times*. However, Ed Garvey did say the senator sat on a committee that oversaw activities of the EPA. He agreed to inform his boss about this problem after he had a chance to read the information I had provided. I said I'd stay in touch to see what action could be taken. We shook hands and I parted for the trip back to Pennsylvania.

As I drove back home fighting rush-hour traffic and singing more sixties' songs, I felt satisfied with my first lobbying venture in

Washington. It seemed so big time.

Sidney and I discussed the results of my visit. She thought I might get a better reception if Linda, our scientific consultant, accompanied me on my next lobbying venture. Linda traveled to Washington often on business, and had an apartment just outside the city. We believed Durenberger to be our best chance for help.

Since Sidney knew Durenberger's top aide, Bert McKasy, I took Sidney's suggestion and asked Linda to accompany me to meet him. Sidney offered to pave the way by calling Bert and explaining the purpose of our visit. She did, and he asked for a summary of the problem and information on my background. (It's possible my radical reputation preceded me via a BFI conduit named Irv.) Just recently, Sidney showed me the letter she sent and I chuckled. She referred to me as hopelessly non-political. It's true. I despise political games and avoid them.

Linda and I set a date, time, and place to meet for lunch outside Washington, after which we'd go together to Durenberger's office. While we ate at one of those restaurants featuring an all-you- can-eat buffet for a set price, we caught up on each other's news. We feasted and talked, and laughed and talked until we looked at our watches. With all the time we had earlier, we were now running late, so we paid our bill and dashed out the door.

When we got near the Russell Senate Building, we had to ask for directions. (Just because I'd been there before didn't mean I could find my way a second time.) A well dressed gentleman directed us to take the underground train professional lobbyists and government officials use. We raced through the tunnel like two laughing lunatics, nearly knocking people over in our path, then jumped on the miniature train. We had great fun. We reached the building, asked for more directions to Durenberger's office, and got there only a little late.

Bert McKasy was delightful. He welcomed us to sit down across a small coffee table from him and tell our story. With adrenaline running high after our little jog, we gushed forth the history, geology, biology, geography, hydrology, and chemistry of BFI's proposed Flying Cloud Landfill expansion, also mentioning a few examples of BFI's dirty tricks. Linda and I both took turns relaying the information, chiming into each other's conversation when we had something important to add. Bert appeared astonished with the situation and laughed, a true indicator he was an intelligent and honest man. He implied the whole story sounded completely ludicrous. We insisted we spoke the truth.

"What is the Minnesota government doing about this situation?" Bert asked.

"Oh, nothing," I replied, until a school bus full of mothers blockaded BFI's landfill entrance, and citizens paraded to the State Capitol with coffins, and we marched two miles around the dump with hundreds of people, and so on. In fact, we pointed out, the Metropolitan Council and MPCA had recommended the expansion as an acceptable solution for the county's waste.

Bert agreed to ask Senator Durenberger to write a letter to EPA chief William K. Reilly on our behalf, which the senator did. Later, we received a fairly generic reply from an EPA underling, informing us the principal responsibility for the situation rests with the MPCA. The EPA official called the Minnesota agency and asked for the expansion's status as if we had no inkling what was going on, and reported it to us. Then he gave us an MPCA phone number where we could obtain more information. He should have asked us for the number; we knew it by heart.

Pilgrimage to Love Canal

The summer of 1989 began quietly. Everyone in Eden Prairie slept easily knowing the landfill was shut down, and the expansion case would be tried in court on issues we had a chance to win. Lawyers from both sides contacted witnesses and prepared them for a fall trial. My phone bill dropped a little.

During July, Ivan had a business trip to Canada, so the children and I decided to tag along for a vacation. En route, we planned to stop at Niagara Falls to sightsee. I remembered the area is also the location of Love Canal. When I looked on the map, naturally, the disaster site wasn't marked; I'm sure New York officials would as soon forget it ever happened. So we decided to ask for directions when we arrived in the vicinity.

As I recall, we drove north to the Canadian border with Paul and Neil; Pierre was visiting Grandmaman in Belgium. We arrived at Niagara Falls the morning of the second day, and began searching for Love Canal. A gas station attendant directed us to the neighborhood. Once we found it, I asked another man getting out of his car, the exact location of the site. He looked at me with sad eyes and pointed the way, explaining he formerly lived over the canal. I inquired whether his family experienced any problems. "My two sons are dead," he said. "I don't know whether it was the chemicals or not." He declined to answer any further questions and walked away. I retreated to our car, realizing I had just met a casualty of the infamous Love Canal.

We drove a few more blocks, parked, and walked the streets of a once-thriving community where children played. Today it's a virtual ghost town. There's an eerie silence, a strange calm haunting the neighborhood. Homes are boarded up or bulldozed over, as in the case of Lois' house, and untrimmed shrubs and trees cover many windows and doors. However, the lawns were all mowed. I suppose the city cuts the grass to avoid an appearance of total decay.

In a central area surrounded by a high, chain-link fence, signs hung at intervals read: "Danger! Hazardous Waste Area! Keep Out!" Behind the fence, enormous pumps worked night and day cleaning up the still contaminated groundwater. Just a few years ago, a park and elementary school stood on this same land.

It's difficult to imagine this was a killing ground. The sky was blue, and only a faint smell of chemicals tainted the otherwise pristine air. But most of us know the story that made headline news around the world: how 22,000 tons of buried hazardous chemicals seeped through the ground, sickening nearby residents, and killing or disabling some of them. This was the first time in history we experienced this type of mass human wreckage from illegally dumped toxic wastes. Lois Gibbs, then just a 28-year-old mother with only a high school education, forced the United States' government to rescue her family and neighbors from this horror. Today, "Remember Love Canal!" is a battle cry for those forced to live with buried poisonous chemicals in their backyards. Never, never, should this happen again, I concluded. And never on land named after the Garden of Eden.

THE POLLUTERS

11 | THE PAYOFF

O ur smugness over recent victories in the dump case screeched to an abrupt halt after my return from Canada. Scott called to tell me Judge Klein ordered both sides to sit down and negotiate a settlement.

I couldn't believe it! Not again. After seven and a half years of agonizing work, we had to deal with this same problem all over again, the one, years earlier, that forced me to get involved initially. Of course it made sense. BFI waited until the new HHHA leadership floundered, then moved in for the kill; history was repeating itself. By God, they wouldn't get away with this, I angrily decided.

Very, very upset, I called Will Collette. With his strong voice, he calmed me down and said, "Tell Scott to make BFI an offer they're bound to refuse, something laughable, like a 1% expansion, to stay at the table. Don't allow the city to negotiate behind closed doors without a representative of the citizens there," Will said. I felt somewhat reassured after our phone conversation, and relayed this message to Scott.

I wasn't sure Scott understood me, though. He seemed relieved to negotiate a settlement. I kept saying, yes Scott, but put something on the table BFI will reject. Yes, yes, I hear you, he said. However, I had my doubts.

These days, life at the Anderson household was hectic. Recently, a baby arrived on the scene, and both Mary and Scott worked. How they could juggle family, jobs, and the dump was more than I could fathom. They must be tired, I reasoned, and they were.

Judge Klein later explained to me he decided to explore a possible settlement because he believed the county needed more landfill space, but he could see problems with BFI's expansion. Klein said putting more garbage on top of old garbage in the unlined horizontal expansion made little sense, since the dump was already badly leaking. And after all the testimony he listened to in court, he didn't see a clear winner, because scientists from both sides swore their client's position was technically sound. Lastly, he worried that if BFI lost the case, there might be no clean-up of the site. Personally, I wondered if the prospect of a long, ugly trial involving the issue of BFI's integrity might have prompted Klein to seek an alternate solution to resolve the case.

Over at Eden Prairie City Hall, the city's lawyer, Ric Rosow, told the city council in a special, secret session that he had only a 30% chance to win the case. Our attorney, Grant Merritt, wasn't invited to this meeting, nor consulted on these odds. The council decided to send their colleague, Dick Anderson, to bring back a settlement for a vote. As these negotiations proceeded under the direction of Judge Rhea, I worried, but kept reassuring myself the city and its citizens were just following Klein's orders. I couldn't imagine either group would cave in after all the effort we consecrated to defeating the expansion. How little I understood human frailty when money dangles within reach!

Grant called Scott and asked him to arrange a meeting of citizens to draft a package "you can live with." Jerri volunteered to hold it in her poolside gazebo. Grant discussed the settlement talks, told the committee to get something on paper, then left. At this point, Scott told the group I said to negotiate a settlement. Most everyone agreed it was the thing to do. Jerri, however, said, "I can't believe Susan would say a thing like that." She stood alone in her refusal to accept a settlement. (Cary Cooper, who arrived late, supported Jerri's position.) Scott insisted I said to negotiate a deal. Jerri decided to see if it was true, walked back to the house, and telephoned me.

When Jerri told me what Scott said, I was angry he misrepresented my position. I thought I had made myself clear: put down numbers BFI will quickly reject. Jerri was livid on learning this, and stomped back to the meeting. She repeated my words and realized if the fight were to continue, she must take over. She declared to the group, "I'm in charge now."

Jerri and Betsy telephoned the following weekend, very upset. Apparently, Scott called the mayor to tell him citizens were 50:50 for a settlement. He made the call without conferring with the others. Worse yet, the group was tired fighting and ready to give up. I talked a long time with Jerri and Betsy, trying to reassure them we could work through these difficulties. I volunteered to fax my position to the committee to clear up any misunderstandings. I had also scheduled a trip to Eden Prairie to attend the Pidcock wedding the following week, and told them to "Just hang on;" I'd be there shortly.

I immediately faxed Grant the details of my conversation with Jerri and Betsy. In bold letters I scribbled across the page: "Please Destroy When Read." Then I faxed off a two-page letter to Environmental Committee members applauding their year's success and urging them on: "You bear the truth and eventually it always wins. So keep a positive attitude, take courage, and press on."

Ivan decided to take a week off from work to vacation with the boys, while I had fun with friends in my beloved Eden Prairie. Jerri set the week aside to work with me on dump matters. It remained to be seen how much fun I'd have flying off to battle once more.

What a wonderful wedding those Pidcock's throw! Live chamber music welcomed guests to Pax Christi Church, dancing French girls entertained us at the reception, delicious food satisfied our palates, and lively debate, provided by local politicians, engaged our intellects.

I managed to hitch a ride to the reception with Dick Anderson, the city's chief settlement negotiator, affording me a private, one-on-one conversation with him. As we drove along, I argued how an expansion could affect the health of nearby children, cause home values to crash, and so on. Nothing I said budged him from his dug-in position.

Was this the same man who walked in our "Ring Around The Dump" protest June 1, 1986? The man who stood before the MPCA

Board, hammering them for failing to protect the citizens? Why the 180-degree flip? I think those questions can be answered in a wonderful little song Liza Minnelli sings in the musical, *Cabaret*: "Money makes the world go round . . . " The city stood to reap millions of dollars in revenues if the expansion went through, which would pay for much needed schools and parks.

Dick gave me one ray of hope, though. He said, "In the end, the people must decide if they can live with an expansion."

At the wedding reception, I joined the politician's circle to discuss the cons of a landfill expansion, while balancing a plate of appetizers and a glass of wine. It was rough going. Ric had them all convinced they'd lose. I pleaded, begged, and sang the dump expansion blues to convince them to back off the settlement talks.

For the entire dinner, I spoke about nothing else to City Council Member Dr. Jean Harris. The poor woman made the mistake of sitting next to me. She listened and made no comments. I regret to say I was terribly distracted at this otherwise lovely affair.

The next day, I sat down with Jerri and relayed the news. We were both depressed. What could we do? Next, I tried telephoning everyone I thought could help, but was put off. Then Jerri and I decided to appeal directly to the mayor. Mark Thompson, Jerri's neighbor, agreed to come with us and put some male teeth into it. We all met at Baker's Square and attempted to persuade the mayor to turn down a settlement. Nothing worked. He played back the same warped record I had heard in 1985: only a handful of hysterical housewives in the Homeward Hills area are concerned about the expansion; most people in Eden Prairie don't care; tra-la-la. He would live to regret that song.

The situation was serious.

The city was ready to deal: dollars for a dump. They were probably tired of the expense which now totaled close to a million dollars, fed up with monitoring the process which gave them endless headaches, anxious for the extra revenue an expansion would bring in, and believed HHHA leadership was weak. On this last assumption, they were correct.

Scott and Mary Anderson could no longer lead; family and work overwhelmed them. Barb Bohn's husband squashed her involvement. Betsy rejected the job offer, and Jerri Coller was only beginning to sort through the boxes of files she took from Scott. The above circumstances formed a juicy, ripe plum ready for BFI pickers.

Jerri, Mark, and I brainstormed together and decided to watch the city's activities carefully. Mark handed me his fax number to use night and day, and we pledged to fight this settlement. In a matter of hours or days, all could be lost once again.

Before I left Eden Prairie, Dick Coller privately told me Jerri wasn't a leader type like me. Perhaps he was concerned his wife had taken on more responsibility than she could handle and might fail or give up. I don't know the reason for his conversation, but I vividly recall his concerned remarks. Time proved Dick wrong. Within several months, Jerri lit a bonfire under this city's conscience that nearly burned the council alive.

Ivan didn't pick up a happy wife at Harrisburg Airport in August, 1989. After his generous offer to send me off to have fun, I returned glum. I realized I could no longer influence events in Eden Prairie from Pennsylvania without a strong leader at the helm of the association. Would Jerri be the one? Grant questioned me on the phone about Jerri's capabilities. Until stepping into the ring and fighting, I said, people never know their strengths. Look at me! My brothers laughed when I told them I blockaded a landfill; they believed I was incapable of such radical behavior. But circumstances pushed me to do the unthinkable. Father Tim once said, the strength of our will determines our success. I knew Jerri dearly loved her children and wanted to protect them from harm. That was motivation enough. Her skills were probably greater than mine when I began. The rest, she can learn, I told Grant.

I now directed my HHHA correspondence to the Coller household via Mark Thompson's fax machine to keep phone bills under control. But I still kept vigil by the phone for the next few weeks, speaking with Jerri, Dick, Mark Thompson, and Grant to plow through these difficulties. I also wrote and called everyone I thought might influence the situation. Ric Rosow received my stinging criticism for leading the city council down the let's-make-a-deal path. And each council member received a letter urging abandonment of settlement talks and continuing the contested case hearing. I lost sleep again worrying about the outcome; I relived the same nightmare I had experienced in 1985.

The negotiating team met in closed door sessions in the Flour Exchange Building Dick Coller later referred to as smoke- filled back rooms. The group was sworn to secrecy. Jerri sat in on negotiations every day. With her young children home for summer vacation, her weekly daycare bill exceeded $60.00 for Monday through Thursday, 9:30 a.m. to 2:30 p.m. sessions. Grant was in a very difficult position. He represented city officials sitting at one end of the table, dealing for an expansion, and citizens who sat at the other end, vehemently opposing it. So he just said nothing and took up space at the center of the table, anxiously observing the proceedings.

Jerri told me she felt helpless as the city and BFI bargained away the future of the neighborhood, and the health and safety of children. To remind the negotiating group of what was at stake, Jerri neatly set 8 x 10 photos of her children on the table, facing the Judge. As weeks wore on, the settlement firmed up. In desperation, and with tears running down her face, Jerri grabbed the pictures of her children and interrupted the group's concentration with an outburst of emotion.

"What about our kids? What will this do to our children? You're talking about our children!"

Judge Rhea took Jerri aside hoping to calm her down but Jerri felt the judge didn't understand the magnitude of the problem and lacked empathy. This was more than just a job to her.

By mid-September, talks reached a critical stage. Ric Rosow, attorney for the City of Eden Prairie said to BFI, "If we get what we want, you have a deal." Jerri, very upset, went home and broke her vow of silence to tell Mary Kennedy what happened. That evening, Mary telephoned me in Pennsylvania and said, "Get over here tonight. The city is ready to sign an agreement."

I didn't make any promises; I wanted to speak with my family first. We talked it over and Ivan and the boys felt I should return. "After all the work you've done, Mom, you don't want to let this expansion go through," said Pierre. Before I made plans, I decided to call Jerri and see how desperate the situation really was. Jerri felt they could handle it themselves. I pledged my full support and help if they needed it.

It was a good decision. I couldn't run back for every crisis. This was their fight now, and eventually they had to search their souls, solve the problems, and lead the citizens themselves. With perhaps a little direction and encouragement, I knew they could dig their way out.

Jerri passed a torturous weekend. What could she do to stop the city from signing the nearly completed settlement? Should she do anything? She talked it over with her husband, Dick, and he said, "Will you be content with a settlement after all the work you've done? I can't make that decision for you. You have to do what you feel is right in your heart."

Monday, the final wording for the agreement was worked out.

Then, Leslie Davis blew the whistle on Channel 11. He accused the City of Eden Prairie of "selling out." The news station called Jerri to substantiate Leslie's claims. "All the blood drained from my body when I heard what Leslie had done," Jerri said. She told the reporter, "I'm sorry, I can't discuss this," and hung up. The press persistently plied her for information. Jerri finally relented Tuesday morning and allowed Rick Cupcella to come out to her house to discuss the situation. Betsy came over to lend moral support. Jerri said she was "so nervous about every aspect of the interview," but Rick was very encouraging. He asked to see the dump and talk up there.

Jerri drove the family van to the landfill and proceeded to get a flat tire on enemy territory requiring the services of a tow truck to remove it. Not a great beginning. As they prepared for the interview, Betsy motioned to Jerri to button up her blouse. She didn't want her friend looking sleazy on TV. As Rick and Jerri walked along the bluff next to the dump and talked, cameras rolled. They discussed the background of the case, then Jerri explained the city was ready to sell out for a big cash settlement. I saw the taped interview a few months later and for a novice, Jerri did exceptionally well.

That afternoon, Jerri telephoned Grant and said, "Watch the evening news."

"What did you do?" asked Grant.

"You'll see; bye!"

Next morning, after seeing Jerri featured on the 9:00 o'clock news, the Judge called a special meeting. Jerri, Alex, and Barb Bohn attended. As Jerri walked into the room, "They all gave me dirty looks except Grant," she said. (I imagine Grant's heart must have been singing, "Hallelujah! Get us out of these negotiations, Jerri.")

The Judge reprimanded the two whistle blower, then discussed their fate. A big debate ensued. Leslie was banned from the talks forever. No one objected. Then, there was that problem woman, Mrs. Coller. Grant

229

argued she should be allowed to stay. Judge Rhea asked her to leave the room so they could decide what to do with her.

When Jerri re-entered the room, the Judge told her to sign a statement forbidding her to openly discuss the settlement talks. "Breaking this agreement could cost you $5,000 or jail," warned the Judge.

I guess I'll go to jail, thought Jerri. (She knew full well there was no money in the association's account.)

Judge Rhea then gave Jerri a statement to read to the press which waited in droves outside the building. She complied. When Jerri finished, a reporter demanded, "Did they tell you to shut up?" Jerri glanced at Grant with downcast eyes, and he nodded in agreement. So she looked back at the reporter and replied, "Yes!"

Jerri returned to the meeting room where further refinement of the settlement language took place. Her resolve to uphold the secrecy pact was wearing thin. She announced, "It's just not right to negotiate for thousands of people without their consent."

The press sniffed a good story. They pestered Jerri relentlessly for the settlement terms. They sensed she could be coerced to talk. Finally, Jerri held a press conference on her front lawn and said, "If they (city) accept this plan, they're indeed being bought." Then she spilled the details of the buyout. The city stood to gain 39 million dollars windfall if the landfill was expanded. BFI could reap a billion dollars with a 90% expansion, and would control a monopoly on landfill space in Minnesota. Yes, that's right, 90%. That was the final figure agreed upon during negotiations, after the city began at zero and BFI started at 100.

For people living near the dump, thousands of dollars awaited them if they accepted the package. If you lived less than 1500 feet away from the dump, designated Tier 1, you got $11,000. Had I stayed in Eden Prairie, I would have been one of the chosen 40 to receive this big money.

Those living farther from the dump got fewer thousands. People residing up to 2700 feet away were covered in the buyout. Then BFI offered property value assurance: the Tier 1 group would be given up to $19,000 if their property values fell. What a deal! With an initial payoff of $11,000, plus this $19,000 for property devaluation, the Tier 1 group had a grand total of $30,000 with which to buy another home if theirs became worthless. To my knowledge, the only house in Eden Prairie you can purchase for $30,000 is for dogs. Furthermore, there was a stipulation in the agreement that no one could hold BFI liable for damage to property and health after they signed. If you had a very sick child, $30,000 would not go very far to pay hospital bills.

After seeing Jerri Coller spill the payoff on television, citizens of Eden Prairie became enraged the city council would consider this back room deal. Angry callers jammed the politician's phone lines with their complaints. As a result, city officials struck a compromise with the people. The mayor announced, "We will hold two public meetings. Turn out people opposing the expansion, and make sure there are more than 200 from all over Eden Prairie to show us this is not just a Homeward Hills concern." Jerri took on the challenge and crossed the Rubicon. Negotiations were suspended.

Jerri Coller hardly resembles anyone's idea of a radical. This cute, blonde, blue-eyed mother of two looks like your classic vulnerable female capable of snapping in a gust of wind. She stands at 5'1 and weighs in at a little over a hundred pounds. Dieting and exercise keep her fit to volunteer for a multitude of school activities involving her children, Richie and Renee. Richie is the same age as my Paul, and Renee entered first grade in 1989.

Like me, Jerri didn't work outside the home when her children were young, so she could give them her full attention. Like me, she obeys the rules of society. And like me, public speaking terrorized her. She told me that in high school her speech notes rattling in her hands were louder than her voice. Unlike me, she grew up in northern Michigan and studied business in college.

Jerri understood some of the inside mechanisms of the association, because we worked together since 1987. But there was so much to learn in such a compressed time frame. There were piles of scientific reports to read, a nonprofit corporation to be run, the press to be dealt with, political players and agencies to get to know, and the community to be organized for a good turn out at these meetings. Lastly, she'd have to testify before citizens and government. Unfortunately, Jerri didn't have the benefit of a slow immersion into the sometimes corrupt world of business and politics, as I did. I feared she dove into a raging river in which she might drown.

231

Soon after Jerri first took over the association, she telephoned to brief me on one particularly grueling meeting, and confessed she vomited from the stress. The strain of crusading wreaked havoc on her physically and emotionally, but her spirit stayed strong. This diminutive mother fooled us all. With a torrent of fresh energy, she came forward to lead the citizens of Eden Prairie in this difficult time they'd long remember.

Master's candidate in Environmental Pollution Control

As I shivered in the damp, winter air sorting aluminum cans from bi-metal cans at the Hershey recycling center on Chocolate Avenue one Saturday in 1989, I struck up a conversation with an enthusiastic young woman performing the same task. I asked her what she did for a living. She replied she was studying for a Master's degree in Environmental Pollution Control at Penn State. How wonderful, I exclaimed! Then I proceeded to ask her details of the program, admissions requirements, and directions to the campus.

Soon after, I drove down to the capitol campus of Penn State University located in Harrisburg, adjacent to Three Mile Island (the Varlamoff family seems to have a strong affinity for disaster sites), and met with the department head, Dr. Ezard. He convinced me to put in an application and signed the form to start the admission process. In addition to a copy of my transcript, the university required three letters of recommendation. Sidney Pauly, Grant Merritt, and Joe Mengel obliged. Within days of receiving their letters, Penn State accepted me into the program.

I had several reasons for doing graduate work in this field. First, whatever information I learned in class could be used for the case in Eden Prairie. Second, I would elevate my status from hysterical mother to graduate student, and lastly, I wanted to prepare myself for a career in environmental pollution control.

Eighteen years after completing my undergraduate education, I walked into Solid Waste Management CE 476 and took a seat in the front row, dead center. The graduate program was only a few years old, but the packed class indicated environmental pollution control would be a growth industry in the future.

It pleased me to see I wasn't the oldest student; several others had

232

passed their thirtieth birthday, also. However, I felt certain I must have been the most scared, but most motivated student in class. I wondered whether I had the concentration and intelligence to succeed in graduate school at my age. Or, would I discover I had a premature case of Alzheimer's disease and be forced to quit? I worried, too, I might not find time to study between car pooling, PTA meetings, working part time at the school to pay for my tuition, and entertaining Ivan's business associates. However, I was determined to give it a try anyway.

One evening, our instructor, who worked for a prestigious firm that designed landfills, distributed diagrams of a state-of-the-art facility and explained the technology. This triple lined operation complete with leachate detection and collection systems bore little resemblance to BFI's expansion heralded as Cadillac quality. In fact, BFI's low-tech, single-lined dump with leachate collection system had more in common with a Model T. BFI's proposal couldn't withstand the stringent regulations imposed by many states, including Pennsylvania.

Linda once told me Tricky Dicky participated in writing solid waste regulations for the state. Perhaps, that's why landfill standards in Minnesota were compatible with BFI's budget. I immediately sent off this dump diagram to the Collers. As an engineer, I knew Dick could decipher it and present it to the settlement committee and Judge.

Quickly, I came to enjoy my evening graduate courses. They became the highlight of my week. While everyone else dragged into class exhausted from a full day's work, I arrived all chipper and ready to ask my usual 50 questions. However, I dared not disturb the class too much with my Eden Prairie dump story, lest they get fed up with me.

One evening, after staying late to complete my first exam, I explained the technical aspects of the Flying Cloud Landfill to my professor and asked him if an expansion would be safe for people? He said over the short term it could work, but long term, probably not. The engineered systems could fail, and since subsoils were inadequate to provide a second line of defense against contamination, people would be at risk. Normally, he said, there's a buffer zone of undeveloped land surrounding a landfill.

Naturally, graduate courses require term papers relevant to course work. I chose *The Environmental Effects of Municipal Landfill Gases*, and began calling doctors and scientists all over the country for studies. I

233

also exhausted the Penn State librarians, who helped me search for additional information. One overworked soul commented that the depth of my work resembled a Ph.D thesis, not a master's paper. I explained I hoped to introduce my paper into court as an exhibit in a big environmental case.

Any studies I thought could help Jerri rouse the people, I faxed to her. At the moment, she was working hard promoting the negative impacts of a dump expansion to the community, so they'd attend hearings.

Results of my research startled me. The Department of Energy did an extensive study on nine municipal landfill sites and discovered that ordinary trash dumps produce toxic gases. It appeared that levels of toxins increased even after landfills were shut. Even very low levels of hazardous emissions migrating through the air can cause health problems in the adjacent population.

Dr. Ozonoff published a study conducted in an area surrounding a Boston landfill that received chemical waste. The results indicated that residents living near the dump showed higher than normal respiratory problems, persistent colds, fatigue, and headaches compared to the control group living at a distance from the facility.

A University of California study demonstrated that children are six times more vulnerable to air pollution than adults. Since risk assessment is based on how a chemical affects adult healthy males, a child's well-being isn't considered when a site is evaluated.

Another study discussed effects of mixtures of organic solvents like those found at landfills. Dizziness, headaches, nausea, motor incoordination, and weakness of extremities were problems reported by adults exposed to solvents.

Lastly, a report showed methane gas replaces oxygen in soil and kills off vegetation. As I recall, only dead trees stood on BFI property, and Dwain Warner remained convinced vegetation on the landfill slope was thinning, too.

My paper earned an "A."

Grassroots convention

During the week of August 15, a letter arrived from Lois Gibbs announcing "in recognition for your work, and your willingness to pass it on to others, you have been nominated to receive the acclaim of the Grassroots Movement at Grassroots Convention 89, which is being held on October 6-9, at the Crown Plaza Holiday Inn in Arlington, VA." I was surprised. I didn't know The Citizen's Clearinghouse For Hazardous Waste gave awards, and with all the people across the country fighting for environmental justice, I hardly expected to be one of those commended.

In addition to the awards ceremony, the convention planned many informative workshops. Grant received an invitation to attend the convention also, and we decided to meet there to make contacts and gather information. Since we'd be just outside Washington, we also wanted to speak with the Director of Municipal Solid Waste at the EPA, thinking a personal appeal might convince the agency to take action. Grant offered to set up the appointment.

My college roommate, Rosemary, lived in a Washington suburb, so I called to ask if I might bunk at her townhouse during the convention. Of course, she agreed, and threatened to stage a late night pajama party reminiscent of our college years.

On October 6, map in hand and oldies blaring on the car radio, I left for Washington again. The road was getting familiar, so my chances of getting lost diminished considerably. I met Grant at the convention hotel, and we took a cab to the EPA for our arranged meeting. Before entering the building, we grabbed a bag lunch from a street vendor, so we'd arrive for our appointment well ahead of time. When we arrived, we announced ourselves to the unattractive, overweight secretary. She asked us to be seated while she checked Mr. Solid Waste Management's appointment book.

Grant and I settled ourselves uncomfortably on folding chairs and waited. What a wreck this place was! It was worse than Minnesota Pollution Control Agency and Minnesota Health Department offices. The carpet was worn thin, walls desperately needed paint, and many Sears folding chairs were haphazardly stacked. If this dump was where the EPA's top garbage man worked, then what could one expect in the offices of his subordinates? A cesspool? Clearly, environmental pollution control wasn't a big budget item in Washington, either. That explained why citizens were hitting the streets in record numbers, taking matters into their own hands.

As we quietly mocked our dilapidated surroundings, the secretary said she had no appointment for us with Mr. Solid Waste Management. He was attending another meeting. We argued that this visit was set well in advance and reconfirmed. Such disorder engulfing us, this error wasn't surprising. We requested the estimated time of arrival of our man, and the secretary said she expected him shortly. While we waited, we emptied the contents of our brown bags on our laps and munched on lunch.

By and by, a tall, harried man dashed into the adjacent office carrying stacks of papers. That's him, the secretary pointed out. We hastily put away half eaten sandwiches into our crumpled bags and walked in after him, explaining our dilemma. He said he had no knowledge of his appointment with us; his secretary handles those arrangements. And since he had more meetings all afternoon, he couldn't speak to us. But we could talk with his aide, he suggested. Then Mr. Solid Waste left the way he came, in a whirlwind.

The aide took considerable time discussing the proposed landfill siting regulations, and gave us the documents that were undergoing approval. No, he had no idea how long the process would take before proposed regulations became law, he said. Maybe a year?
Since prospects for EPA help looked poor, we figured that for the grassroots convention, we better acquire the tools to do the job ourselves.

After a quiet dinner on a restaurant terrace in Washington, we registered at the convention, greeted Lois and Will, and attended an evening affair that included singing, an auction, and story telling. The hotel was overrun with a rainbow coalition of more than a thousand environmental activists. Rich and poor, well educated and poorly schooled, and students and grandparents sat together swapping war stories. I even recognized some Greenpeace people who led the bus blockade. Environmental injustice favors no socio-economic group, but it did seem there was a disproportionate number of low income people in the crowd.

The mood Friday night was extremely upbeat. People drew hope and strength from each other as they shared their successes and failures. At the celebrity auction, a book from Meryl Streep sat alongside a handmade quilt presented on behalf of a deceased leader from the rural South. It was enormously touching. I had the impression we'd ride an

236

emotional roller coaster during the next two days.

The following morning, Grant and I studied various workshop offerings and selected six from among 20 possibilities. We decided to divide and conquer; Grant went to lawyer-related groups and I headed first to "Health Effects From Toxic Sites." By the time I got to the lecture room, all seats were taken, so I stood along the walls with many others, mostly women. Finally, when that space was occupied, the door was shut.

Dr. Beverly Paigan, the doctor who forced the government to acknowledge health problems at Love Canal, was our speaker. She began by asking us to identify ourselves, and our reasons for being there. This was more than I had prepared for.

Person after person stood up and recounted tales of sickness and death in their communities. A young Hispanic woman who looked like a teenager, said she lived in government-subsidized housing on a hazardous-waste landfill. In her broken English, she said children in her neighborhood were getting sick; they desperately needed help right away. Another mother collapsed in her seat crying after she told the group she lived near a landfill and her daughter died of leukemia.

An old man announced he had cancer he believed resulted from toxins coming off a hazardous waste site next to his home. Several other women discussed their chemical sensitivity, a condition affecting the nervous system. They said they had difficulty focusing and raising their arms. Their eyes were dull and their skin sallow. A middle aged woman described despair that overtook many sick and dying neighbors near a toxic site, prompting seven of them to commit suicide. A rural doctor wanted to know how to treat victims of toxic poisoning. Such testimony continued for an hour. Much of it consisted of case studies not yet documented in medical and scientific journals; the people little more than guinea pigs and victims of man's reckless disposal of chemicals. The outcome is very clear: as we've poisoned Earth's elements, we've poisoned ourselves, especially the youngest and weakest among us.

The raw courage of people who, despite their own afflictions, continue pressuring government for pollution site clean-up and relocation of victims, restores my belief that humankind can win its worldwide, anti-pollution war. However, battles to protect our planet are obviously not being waged in board rooms, political arenas, or

237

government offices, but in living rooms, church basements, and on streets. Warriors waging them are our Earth's heros and heroines.

Sunday, October 9, Ivan and Paul drove to Washington and joined Grant and I for the awards luncheon and ceremony. We battled crowds for a place close to the front of the room, near the head table where Ralph Nader was sitting. Again I forgot to eat because I became preoccupied trying to get Ralph Nader's attention during the meal. I had heard he was scheduled to be in Minneapolis soon, and I wanted to ask him if he could help the homeowner's association.

When Ralph sat down at the head table, I dashed over to him, explained the dump situation, and asked if there was anything he could do for the Eden Prairie people when he was in their area. He said there was really nothing he could do in such a short time. How about at least a photo with the residents for the newspaper, I suggested. Then the political powers in Minnesota would see that Ralph Nader, the country's most famous activist, supported the Homeward Hills Homeowners Association. He agreed to grant this small favor.

Ralph Nader is a tall, thin, intense man about 55 years old. Harvard declared him the man of the century for his dedication to consumer protection. His speech was simple; no rousing call to arms, just a quiet discussion on environmental problems and solutions.

"How many of you have wind up watches?" he asked.

About ten hands went up.

"It's tough turning that knob every night to reset it, isn't it?"he suggested mockingly.

We laughed knowing full well the batteries from our automatic watches contributed toxic chemicals to landfill leachate; we had no excuse for not using the manual variety.

He proclaimed the Citizen's Clearinghouse For Hazardous Waste (CCHW) as the only effective environmental organization in the country. The audience shared the opinion and thundered an ovation. The crowded room of people who won environmental battles in their communities confirmed his pronouncement.

Then Will got up to conduct the award's ceremony. He called out the names of each state beginning alphabetically with Alabama. Will delivered a fast paced performance, booming out the triumphs of award recipients as they walked to the stage to accept their   certificate from

Ralph and hug from Lois. He took us on the roller coaster ride I envisioned. Collectively, we cheered, wept, laughed, and jumped to our feet for ovations. It was both exhilarating and exhausting; a grand finale for a moving weekend.

Stories slid together as Will told them without pausing. An American Indian stood ramrod straight to accept an award for refusing to allow a dump on his tribe's reservation. There was a woman who shut down a tire factory spewing toxic fumes across her neighborhood, and an African American man who opposed a landfill in his neighborhood for ten years. An old man with cancer walked slowly to the podium with his elderly wife to be commended for his leadership efforts. Next, a young mother who shut the town dump became mayor and ordered the dump company to clean up. Eventually, I heard Will shout Minnesota.

"In Eden Prairie, a group of mothers became so frustrated when the government refused to shut down the leaking dump near their homes, that they did it themselves in a school bus. They didn't know what to do with their kids, so they brought them along." The crowd roared its applause, Ralph Nader said, "Congratulations!" Lois embraced me, and my husband, Ivan, snapped photos.

It was a humbling experience. I felt terribly small in a room so full of brave people, but enormously proud to represent the victory of the Eden Prairie, Minnesota, Homeward Hills Homeowner's Association.

And the wall came tumbling down

November 10, 1989, the world sat stupefied in front of television sets, watching young German men hammer down the Berlin Wall. This barrier that physically and ideologically divided Europe for forty years lay in pieces on the streets by morning. We all rejoiced. The cold war ended, and Communism was mortally wounded.

It happened so quickly, we all wondered why the world permitted a totalitarian system to suppress such a large population of people for so many years. Dissidents spent decades alerting the West to the "evil empire," but in the end, it was the people themselves who toppled the "Iron Curtain" wall in Germany, cut barbed wire in Hungary, and threw off Communist shackles in Poland.

In Eden Prairie, Minnesota, and Hummelstown, Pennsylvania, Jerri Coller and I watched with rapt attention this marvel of our modern world. The news empowered us. We both had the same thought: "If the people could tear down the Berlin Wall, then certainly we could stop BFI from expanding its dump."

During the destruction of the Berlin Wall, Ivan was coaching Pierre's soccer team and noticed one boy who always stood aside from the group. He walked over to speak with him and learned the boy's father was a Solidarity leader in Poland and his mother worked at the Hershey medical center supporting their two children alone. The father was due to arrive soon to spend some time with the family. I quickly contacted Magda, the mother, and invited the whole family to our home for a Thanksgiving meal in November. Then I called my father to ask him if he'd like to join us; his parents were Hungarian, so I thought he might enjoy hearing a true story. He agreed and drove over.

In my mother's fashion, as described by my father, I prepared "enough food for an army." Ten of us sat down to a table laden with stuffed turkey, two fruit pies, potatoes, two more vegetables, and cranberry sauce, complimented by California wine, flowers, and candlelight.

In the soft glow of the room, I studied the face of Gizegovz, who sat directly to my left, thoughtfully telling his life's story in heavily accented English. He was young, perhaps 35 years old. His skin appeared unlined and pale, probably due to his years in prison. And there wasn't much flesh on his small frame, either, probably for the same reason. However, his eyes reflected an intelligence and serenity one rarely sees in anyone, much less in one so young.

After completing his Ph.D. in nuclear physics, Gizegovz explained he worked for the Communist Government earning $30 per month. Magda, his wife, a genetic engineer, made the same pay. Life was a constant struggle for them, with no promise of anything more for the future. They lived in a cramped, three room apartment with barely enough food to eat. Their daughter said the meat on our table cost more that her father earned in one month. As young parents, the futility of their lives, which held nothing better for their children's future, motivated them to join the newly formed Solidarity movement.

"There were 50 men who ran the organization," said Gizegovz. "I worked the underground printing press which printed leaflets we distributed to millions of Poles. The Catholic priests encouraged the revolution. We planned our strategy in church basements, after we attended Mass. The Pope often came to speak with us."

"You know the Pope?" I exclaimed.

"Yes! Sometimes we were so discouraged that without the support of John Paul, I don't know if we could have held out."

I felt privy to information I believed few people knew. A year later, the press reported this furtive alliance to an astonished world.

"Do you know Lech Walensa and Vaclav Havel, too?" I asked.

"Yes, we were all in jail together. The government made the mistake of putting all the Solidarity leaders in the same prison. We just continued our work in prison," said Gizegovz.

"I cried every time I visited my husband," Magda interjected. "It was an awful life, but everyone helped me. While I worked, friends cared for my children, and there was always food brought in."

"So how did you finally overthrow Communism? Tell me the secret," I asked with anticipation.

"For years, we passed information from one person to another until nearly everyone joined Solidarity. It was incredible; farmers, students, scientists, factory workers, mothers, doctors, and, of course, the priests worked together," said Gizegovz. "Once we had the people, we organized citizen protests, and, little by little, wore down the government. It took fifteen years."

I was ecstatic hearing this strategy. This is exactly what we were doing in Eden Prairie, leading a minor revolution. There was no question, we could do it. If people succeeded in Germany, Hungary, and Poland, we could win in Eden Prairie.

Quickly I explained my landfill saga and how we were fighting a billion dollar company to shut down a leaking, toxic dump in my former neighborhood. "Yes, you'll do it," said my Polish acquaintance. "And when you finish, come to Poland and help us. We have the world's worst pollution, because outdated Communist factories have no pollution control equipment. Many people living near them are very sick, especially children."

We sat at our table talking for three and a half hours while candles burned and everyone satisfied intellects and appetites. Except for a few bare bones, I noticed our Polish guests wiped their plates clean. I was willing to bet big money this was one of the biggest meals they had ever eaten. They deserved it. It was our Thanksgiving to them for dedicating years of their lives to help end the tyranny of Communism in the world. I decided to carry their inspirational story back to Eden Prairie.

The people speak out

As the revolution in Eastern Europe ebbed, the one in Eden Prairie gained momentum. Since the unwelcome news of BFI's proposed settlement reached city homes October 19, 1989, the Coller phone rang non-stop from 7:00 a.m. to 11:00 p.m. As a result, Jerri established a command center in her living room to direct citizen-led opposition. The *Eden Prairie News* did a front page feature story on her preparing a public relations campaign.

Everyone who called the Collers was encouraged to attend an organizational meeting at Pax Christi Church with friends and neighbors. Cardboard signs announcing the event were strategically placed in developments along Homeward Hills Road. And Pax Christi agreed to open its Dorothy Day Social Center for the group.

"We never knew how many people would come to meetings," Dick said. "Jerri and I always agonized, waiting to see the turnout. That was the worst part."

"Up until ten minutes before the meeting began, only ten people were there," Jerri told me. "Then within minutes, lines of people formed through the vestibule to sign the volunteer sheet. We filled four sheets! 350 people came! We didn't have enough space in the Dorothy Day Social Hall, so the janitor opened part of the church to accommodate all the people," Jerri said. "We were flabbergasted."

Dick challenged his audience to get involved and stop BFI, the government, and the city from bullying them into accepting a dump expansion. As a short child growing up on the streets of Philadelphia, Dick knew how it felt to be the underdog, and learned long ago how to fight back. Adversity was his faithful companion for years, even after he left the neighborhood to study engineering at Drexel University. During his college days, Dick's father died, forcing his son to finance his own education with student loans. Dick Coller grew up to be tough and street wise, making him acutely aware citizens were being trampled in this settlement. After many others backed out of the arena, Dick lowered his horns, and, like a raging bull charged forward to gore BFI's skillful

242

matadors.

People were evidently moved by what Dick and Jerri said, because a man in back of the room stood up, took off his hat, and passed it, proclaiming: "I can't think of anything more important." Brad Lantz, a professional advertising man, came forward to design newspaper advertisements for the campaign, and 125 people volunteered to put flyers and signs in their neighborhoods.

A core strategy group met regularly at the Collers to plan logistics of their attack machine. Mark Thompson, a neighbor and owner of a very successful computer business, pulled up in his Porsche to help out. As predicted, the well-heeled now felt threatened by neighborhood pollution, and willingly pitched in to stop it.

Brad Lantz designed advertisements showing a map of Eden Prairie overlain with concentric circles, each showing miles from the dump. Bold headlines read, "ARE YOU ABOUT TO BE DUMPED ON?" Various effects of an expanded dump listed in fine print were: "toxic pollutants that can travel 4 miles, depressed home values, 1,000 garbage trucks on city streets," and the question, "How far do you live from the dump?"

This advertisement ran in two local papers for three weeks prior to the hearing dates, and one week in the *Star Tribune*. Forty thousand flyers were printed from it. Then 250 yard signs were made, showing the "Just Say No To BFI" logo and dates of the hearings. The "Just Say No" theme hit me when my son came home from school bearing the familiar button showing a circle with a slash through it, adopted for the anti-drug campaign. Jerri made up buttons and T-shirts with the slogan. Lastly, she blew up the Eden Prairie map to instruct volunteers where to deposit flyers, signs, petitions, and lists of politician's phone numbers.

In the beginning of November, the Dumpbuster Brigade struck. Volunteers descended on Eden Prairie homes, shopping malls, churches, grocery stores, and voting polls, leaving in their wake a trial of signs and flyers urging every citizen to join the effort to close the dump. You couldn't go more than a few blocks without being told to attend the hearings. And if you made a trip across town, repetitive yard signs gave you opportunities to memorize hearing dates and times. Jerri wrote me, "It's going to be hard not to know about this."

Local newspapers now thickened with self-described "distressed, offended, appalled, betrayed, and angry" citizens who penned letters to the editor attacking the city and BFI for "insulting our intelligence, considering such an absurd offer, willing to risk my family's future, blatantly lying, and forcing us to face a potential danger." One creative individual even suggested "in the unthinkable event this council accepts BFI's offer...bring in the band and have them play the death march."

Anxious to support the citizens, I faxed two different letters to local newspapers. They were both published. In the one, I chided the Eden Prairie City Council for awarding me a key to the city as the leader of the citizen's group opposing "the increasing scope of hazards of this landfill" (proclamation words), then turning around "to negotiate a landfill settlement and big bucks for city coffers in exchange for any expansion."

Naturally, BFI didn't lay back and submissively take the beating. It fired cannons, too, in the form of a question and answer sheet mailed to all homes, plus full page ads taken out in the papers. "THE FACTS FAVOR AN EXPANSION OF THE FLYING CLOUD LANDFILL," screamed the newspaper headline. I believe this headline was especially directed at our zealous band of mothers who organized dump opposition, and who BFI accused of reacting emotionally rather than rationally to the issue.

During my deposition, Chris questioned me extensively on my understanding of the dictionary definition of fact. How amusing, I thought. Here before me sits a lawyer who twists and turns information to win cases for his client, and perhaps hasn't even taken a high school chemistry class, insinuating the meaning of fact is beyond my comprehension. I've studied biology, chemistry, physics, and mathematics for ten years in high school and college, and worked at Columbia University in scientific research. Miscalculating facts like these cost Browning Ferris Industries dearly.

Webster defines fact as, "A thing known to be true." And yes, we had plenty of those. We shouted them, wrote them, wore them, carried them on signs, and, in desperation screamed them. Is a mother justified in reacting this way when she senses her children's well being is threatened? I think so. And I don't believe she has to make any excuses for her behavior.

There were three people favoring dump expansion who spoke out in the newspapers. They implored the city council to "Give BFI a Chance,"

"Do The Right Thing," and "to make the tough decision, settle with BFI!"

Looking for all the support they could muster, BFI called in its big guns. Steve Keefe of the Metropolitan Council stated "the seven county area will run out of landfill space by 1993 without the expansion of Flying Cloud." And Rod Massey of the PCA said, "It wouldn't be the least desirable site you could find."

Editors of both *The Sailor* and *The Eden Prairie News* were divided in their support. Betsy Dick of *The Sailor* urged the city council to "Reject the landfill expansion proposal," and Mark Weber of the *Eden Prairie News*, disappointingly said, "a strong argument can be made for accepting the negotiated settlement."

In case someone missed this media blitz on the dump, the *Eden Prairie News* held a written debate between the President of BFI, Jerry Lynam, and spokesperson of the Homeward Hills Homeowners Association, Jerri Coller. Jerri implored the city to follow the example of a Chicago community that said, "No dumps, no deals," when Waste Management offered the city $25 million.

"The dump crusade became an obsession for us," Dick told me later. "That's all we do together. Jerri fights this dump till all hours of the night; my sex life is on the rocks."

"At first Dick didn't help out, but I guess he figured he couldn't stop me, so he might as well join in," said Jerri.

The Coller children, like the Varlamoff boys, also tired of seeing their mother speaking constantly into the telephone receiver. Riche begged: "Mama, please hang up the phone."

With Jerri's fervor to recruit people and raise money, she noticed when she approached groups of people they seemed to disperse, cross streets, and turn corners to avoid volunteering more time and money.

The Collers worried about personal risks fighting BFI. Dick wouldn't let Jerri go out alone evenings. An Eden Prairie businessman who called their home swore the phone was tapped.

During negotiations, a skittish Jerri grabbed her friend Betsy by the shirt and said, "If something happens to me, you have to take over. Promise me you'll do this."

"Oh, don't be silly!" laughed Betsy.

"Please, promise me."

"Oh, all right, Mrs. Coller."

Soon after this outburst, Jerri was in her front yard when a suspicious looking car with blackened windows slowed down in front of her home. Jerri threw her daughter and herself on the ground and lay there until the car passed.

On another occasion, when Dick, Jerri, and the children went downtown to City Center to shop and enjoy lunch at the Food Court, Dick panicked because he couldn't find Jerri in the baked potato line where he last saw her. Convinced some ill fate had befallen her, he called police. As it turned out, Jerri switched to the shorter soup line without telling him. They both laughed at the overkill reaction, but remained cautious thereafter.

The city council set hearing dates for November 16 at Pax Christi Church, 28 at Eden Prairie High School, and December 12 at Wooddale Church. These places were chosen because they offered the largest meeting facilities in town. However, the dates competed with Thanksgiving and Christmas holidays, and a major community event. Was this an attempt to discourage attendance?

Jerri and other neighborhood mothers recruited volunteers to call pages of residents in the telephone book to remind them once again to attend the meetings, in case they were blind, illiterate, or out of town the past month and a half and didn't know about them.

The mayor and city council members had no rest from the multitude of angry residents calling their homes day and night protesting the settlement package. A family life was impossible, said Councilman Doug Tenpas, but it helped to know the will of the people.

With everything ready for a good turnout, Jerri prepared her speech. She decided to use good Catholic strategy: GUILT. Recalling the many times the mayor and council promised to fight the expansion to gain the Homeward Hills vote, Jerri chose to refresh their memories. It occurred to Jerri that local newspapers were a good source of direct quotes, so she drove over to their offices with Bonnie Swaim, my ex-neighbor, and searched their files. "Every time we found something good, we screamed, and the whole newspaper staff came in to see what happened," Jerri said.

These ladies copied the best comments of each city council member, and arranged to have them put on separate slides. These quotes would

246

then be displayed on the giant screens over Pax Christi's altar while Jerri shouted them across the church to the council.

Just before the first hearing, arrangements were made to have Jon Grunseth, the Independent Republican candidate for governor hold a news conference at the dump site. 75 placard-carrying residents stood alongside him when he declared, "This landfill is unequivocally abusive to our environment. It is beyond remedial repair and is inherently flawed." He talked about known "hot spots" and called on Governor Rudy Perpich to join him in a non-partisan effort to close the landfill.

As the hearing approached, Jerri realized she had nothing appropriate to wear for opening night at the dump hearings. (A stay-at-home mother's wardrobe usually consists of jeans, shorts, T-shirts, and party dresses for the yearly office bash.) The queen of the dump purchased for her ensemble a black, straight skirt and blue blazer with lace handkerchief. "I wanted a professional look," said Jerri.

Everything was ready. Then the mayor called.

"I have had threatening phone calls, so I ordered police protection for myself, the council, and your family during the hearings," he said.

"Should we be afraid?" Jerri asked.

"I don't know, but there might be trouble, so I want to be prepared."

12 | "CLOSE THE DUMP!"

T he evening of November 16, 1989, at the rear of Pax Christi Church, Father Tim, arms folded, smiled as he strutted like a six foot peacock. He was proud. His church was being used for a purpose he had envisioned back in its planning stages, as a gathering place for the community. Tonight, it was the center of town. People of Eden Prairie would decide the future of the contested case hearing, and probably the fate of the dump.

"How many people do you think are here tonight," asked one curious person.

"About 1,500," Tim happily replied.

"How do you know?" she inquired.

"I'm the pastor. I know how many people this church holds," he said.

As crowds poured in, Jerri and Dick stood crying in each others arms. Their dogged campaign to get the people out worked. The numbers here tonight should show the city council that the citizens had serious concerns about BFI's dump.

249

Hand lettered placards sprung up everywhere in the church, and a group of children conducted its own parade through the pews. Many people signed up to speak under Jerri and Dick's name. But when the sheet was full and the hearing ready to begin, Craig Dawson, an administrator at city hall, snatched it up. Dick Coller wasn't too emotionally overcome to miss the move.

"What are you doing?" Dick demanded.

"We're going to pick people randomly from the list to speak," replied Craig.

"Who the hell is in charge here?" Dick demanded. "Ric Rosow told me the sign-up sheet would be used to determine the order of speakers." Pointing his finger at Craig, Dick ordered, "You come with me."

Dick marched up to Ric with Craig trailing behind and asked him the procedure for the evening.

Ric concurred with Dick; the people would be called in order from the sign-up sheet.

"This was our home turf," Dick explained to me later. "We felt confident holding our first hearing at Pax Christi, because we attended services here, plotted strategy in back rooms, and knew the pastor."

At precisely 7:30, Dick grabbed the altar microphone, and yelled, "CLOSE THE DUMP!" On cue, several other people planted around the room jumped up and chanted with him, "CLOSE THE DUMP!" clap, clap, "CLOSE THE DUMP!" clap, clap, "CLOSE THE DUMP!" clap, clap. People joined in. Like a fundamentalist preacher, Reverend Dick whipped the crowd to a frenzy, repeating the same line 50 times. No one dared stop him. Citizens reveled in the opportunity to shout their frustrations. The roar of their voices was deafening.

After Dick relinquished the microphone, the moderator, an administrative law judge, told the audience, "This a public meeting; this isn't a Timber Wolves game and not a Vikings game. Therefore, we would appreciate it if there is a minimum of hooting, clapping, cheerleading, and the like." (This unfortunate man must have deeply regretted his decision to chair this meeting.)

Ric Rosow followed him with an explanation of settlement terms. Then the first speaker was called, "Jerri Coller."

The room burst into prolonged applause as Jerri walked to the podium. "The people wouldn't stop cheering," Jerri told me. "I didn't know what to do!"

"But I was calm. I remembered what you said, Susan, 'When you speak the truth, there's nothing to fear.' Forces outside me took control and I gave it everything I had," she said.

"Well, here we are. The people BFI called the handful of hysterical housewives. Pretty big handful!

"It is incomprehensible that you (city council) would even consider exchanging for money any degree of expansion which jeopardizes our children's health and our environment."

Vigorously wagging her finger at the city council, Jerri angrily shouted, "Shame on you.

"I tell you now, the people don't want BFI's money any more than they want their dump. Are you listening yet? We are now ironically forced to use the identical arguments against dump expansion that you yourselves used so eloquently in the past to the Metropolitan Council, MPCA, and Attorney General."

The oversized screens over the altar lit up with words from each council member as Jerri spoke them aloud.

"It's my fervent desire, yes, my dream, that we can put all this discord behind us. And by we, I mean a united again city.

"If they can tear down the Berlin Wall, surely we can close the dump." A standing ovation brought the house down .

"Rarely in a person's life do they get the opportunity to make a difference in the world. You have this opportunity. Don't turn your back on it!"

Jerri's speech was interrupted 11 times with applause. Need I say more about the quality of her message and the people's appreciation for her efforts? Susan Schultz, a neighbor, acclaimed it, "the best speech I ever heard."

Dick got up next and identified himself. "I am Dick Coller. I live in Deerfield about a mile from the dump." Dick had an uncanny way of strongly enunciating the beginning and ending consonants of dump to make the word sound vulgar. He also set a standard for the evening. Each speaker after him identified themselves and the distance they lived from the dump.

"Why are so many here? Why are they mad? Why are they angry?" asked Dick. "Because they've been pushed too far, and they're not going to take it anymore. Have you ever seen the neighborhood bully come up to someone and push him? He shoves again and again and you wonder why and when the person shoved will fight back. Some never fight back. They get pushed again, and later the bully continues to take, take, take. But almost everyone will fight back when home and family is threatened."

"The unkindest shove of all? Our best friends decide to join the bully. They make a behind-closed-door deal in a smoke-filled room to join forces with the bad guys for money.

"Yes, we're damn mad, and we're ready to fight back.

"This is our city. We, the people own it, not BFI.

"This billion dollar corporation cannot and will not march here from Houston, Texas, with its big cowboy boots and a bag full of money, and buy our city, buy our quality of life, buy our environment, buy our property values, buy the health of our children. We will not allow it.

"Close the dump! Close the dump! Close the dump!"

The people loved Dick. He had the guts to do and say what most of us felt, but would never attempt. And he did it with such incredible intensity, no one forgot.

Steve Frick, the citizen who gathered hundreds of names on the petition we gave to the Metropolitan Council, stood up next.

"I wrote to Governor Rudy Perpich on December 12th, 1987. I told him I thought it was a crime to pollute a river which is the source of drinking water for people downstream in Minnesota and in other states." (Remember the MPCA and BFI considered the Flying Cloud dump site good, because pollution flowed into the Minnesota River, which empties into the Mississippi.)

"I have in my hand nine certified letter receipts. On Wednesday morning I mailed nine certified letters. These letters were mailed to the governors of the states that obtain their drinking water from the Mississippi."

Terry and Betsy Nigon, parents of three sons, followed. They wanted to avoid repeating the emotional arguments of their friends, the Collers, so they simply stated the facts.

"It is a fact," began Terry, "that the dump sits on the sand. It is a fact . . . ," and another morsel of proven information. He continued in this manner, reciting 15 facts about the polluting dump. I guess he was

252

trying to impress BFI officials we were technically able after all. He finished by saying, "Everybody cannot live upstream. We are responsible for people and wildlife living downstream.We are responsible for our children, our neighborhoods, and our future generations."

The audience clapped enthusiastically for Terry and Betsy, and for each of the other speakers.

As I read the transcript from the meeting and viewed the videotape, I realized how intelligent, clever, and determined these Eden Prairie citizens were. Dick Perry told the council an old French proverb: "Happy people don't make history. I want the city council to know the people in this room tonight are not happy, and we are ready to make history."

Jan Jenkins, a teacher, stated, "As adults, Planet Earth is a gift we have borrowed from our children. Implicit in that statement is, if you borrow something folks, you must give it back, right? In what shape do you want to give Planet Earth back to future generations?"

Lynn Forster, a resident from the other side of town, presented 1,000 signatures to the city council opposing the expansion. She received a standing ovation for her efforts.

Jeannie Helling brought "greetings to Eden Prairie from upstream, Le Sueur County. We have a twenty-year-old landfill too, same as you. State-of-the-art, you know, twenty years ago. It's a little spot in the marsh where they just started throwing stuff."

She expressed her condolences Eden Prairie had "a bad, sand based landfill. We've got the best clay in the state, right in my backyard, and I can't tell you how excited I am about that. But I also want to tell you about our neighbors. They have cows, they have pigs, and they have children. The pigs get sick and fall over and die. The cows are getting bloody noses. Just recently, this poor couple has been trying to figure out if their children have leukemia. They'll be running and all of a sudden they start bleeding out of the nose and can't stop it. And I'm here to tell you, you're not alone. Don't forget, we're upstream."

Mary Kennedy emphatically stated, "We know we elected these people, and we can throw them out of office, too."

Mary Kimitch expressed concern for the expansion's effect on her children, who have asthma. "They've done remarkably well since January, 1988. The dump was closed in March, 1988. For the sake of my children and all children, please don't let the dump reopen."

Tom Schwartz, owner of a property tax reduction firm, told everyone he targeted Zone A, the area closest to the dump, as a Dead Zone. "Six or eight years down the road, these properties will be worthless. Salvage value."

Then a burly, red-haired firefighter, Walter James, held the audience spellbound with his eye witness account of the famous landfill fire. I tried contacting him numerous times, but he never returned my calls. "Wally" came forward this night out of a sense of duty to his city; he said he realized there might be negative repercussions, but with so much on the line, he'd take the chance. Jerri told me, "You could hear a pin drop in the room when he spoke."

"We had to have seven fire departments there for seven days, twenty-four hours a day. I stood in everything. Syringes, needles, aerosol cans, batteries, flammable liquid. A hodge-podge soup of waste. I and every firefighter there knew we were not fighting or standing in clean garbage, but that we were in a hazardous landfill.

"Horrible plumes of dangerous clouds of heavy smoke and gasses drifted over our neighboring city, homes, and families.

"On the third day I was out there, I was overcome with gasses. I don't know what happened to me, but I woke up in the ambulance half way to Methodist Hospital.

"On the fifth day, I saw several hundred 55 gallon drums standing in a black liquid. Well, I would like to know where those 55 gallon drums are today folks, wouldn't you?"

He spoke for some time, giving more details of the fire, then concluded: "Those are the things I saw and experienced. And I'm here to tell you folks, if they open up that landfill again, we're going to be out there again, and I just hope nobody else gets hurt."

The church full of people jumped to their feet and cheered Wally. This brave man became an instant hero.

One small footnote: Wally was diagnosed with cancer sometime after the fires and appears to have recovered. One cannot help thinking maybe fighting fires at the toxic town dump could have precipitated his illness.

Mark Hoaglund spoke after him and challenged Councilman Dick Anderson, his former physical education teacher who taught his students good values and to "have some integrity."

Another man had this to say about the air stripper: "This really stinks!"

And several others proposed alternate names for Eden Prairie if the expansion were approved: "Dump City" and "Garden of Garbage!" !

The room was charged with energy. Eden Prairie citizens unanimously told the city council how they felt about the settlement and the dump: they hated it.

For Jerri, it was a magical night. By the time it was all over, she was the evening's superstar. People enveloped her with their warmth and congratulations. Although I'd love to have witnessed this spectacle, I believed my place was in Pennsylvania. It was important to show city officials the new leaders were in full control of their own destiny. And they deserved to bask in the spotlight alone for their superb performance.

Dick and Jerri telephoned late that evening to tell me the results. "Susan, you would have been proud of me if you had been here," said Jerri, fighting back tears. I was not only proud, I was overjoyed. My intuition about her leadership ability hit the bullseye. And how good to see the work I left behind rise to such a crescendo.

There were rumors the turnout at the second hearing would be low. (Sounds like BFI-inspired propaganda.) Taking no reprieve from her public relations campaign, Jerri made four giant billboards and pounded them into the ground along well traveled roads. She smashed her hand in the process, attributing the black and blue marks to poor eye and hand coordination. A neighbor ingeniously decorated the sign along Homeward Hills Road with Christmas lights to highlight the season's social event. Someone else added a spotlight.

One hundred fifty more yard signs were printed and placed around town. Flyers were sent to people who signed in at the last meeting, and the telephone tree was re-activated to call the entire population of the city again.

No advertising opportunity was neglected. Jerri, her neighbor, and both their children decided to drive out to Graffiti Bridge, a railroad bridge that was a historical landmark, and leave a message. They left, armed with extension ladders and buckets of paint to cover up with their own logo, the initials of Eden Prairie's current lovers and the year the last high school class graduated. Managing to avoid being hit by traffic, and only spilling paint on their clothes, the group scrawled, "SHUT THE

DUMP" across the face of the bridge.

The Environmental Committee of the Homeward Hills Homeowners Association (HHHA) arranged for me to ask the council in person, at the second hearing, why they'd award me the key to the city for shutting the dump, then propose to expand it. Mark Thompson signed over his frequent flyer ticket to me, and I was off again with another speech to Minneapolis, while my wonderful father cared for my boys after school.

I arrived at Brenda's house only hours before the meeting began November 28 1989. She forgot I was coming, or perhaps I neglected to tell her I was coming, so when Brenda came home from work and found me in her living room, she was surprised. "I thought the hearing was next week, she laughed."

"No, it's today, and I've got to run." And off I went.

I hitched a ride with a neighbor for the five-mile trip across town to the high school. I chuckled, seeing the dump billboard twinkling in the night as we rode up Homeward Hills. The town was littered with lawn signs along our route; Jerri had been hard at work. When I exited the car in the high school parking lot, I shook. I'm not sure if it was from cold or stage fright. Joe Mengel stood in the school hallway, so I greeted him and confided my condition. "Oh, you're just excited. Once you get going, you'll be fine," he said.

After I entered the auditorium, I spun in circles trying to absorb the bustle of spontaneous activity around me. Hand-lettered signs shot up everywhere, a coffin was being dragged across the wooden floor for donations, and Mark Thompson paced in front of the bleachers warming up the crowd with "Close the Dump!" cheers shouted through his megaphone. People continuously poured through doorways. Five years ago, a few of us plotted in my living room to shut the dump. Now the room could barely contain the protesters. There was standing room only.

I walked to the podium and signed up to speak to make sure I'd have a time slot. Sitting nearby was BFI lawyer, Chris. I decided to visit with him. Smiling, I patted him on the arm and asked him if he missed me. He laughed and said of course he did. I just wanted to make sure we stayed on good terms. Then I joined the radical HHHA leaders in front row seats on the wooded floor facing the podium.

The excitement in the room was higher than the ceiling, except among the stone-faced men sitting behind us with their arms folded.

They didn't seem to be having any fun. (Later I learned they were undercover policemen assigned to guard us.)

Cheerleader Coller led the evening's program with a pep rally. (Here's a man who wears many hats.) "Close The Dump!," clap, clap. "Close The Dump!" clap, clap. Coller kept shouting the slogan. The delighted crowd responded in rhythm, pounding the wooden bleachers and rocking the room till the rafters shook. I laughed in amazement. I participated in many high school rallies, but this one beat them all. These people wanted to win this game badly; you could feel the determination in the room. I was glad to be part of the action this time, and not hearing it via telephone.

A diminutive woman came forward after the opening cheer and introduced herself as Betty Craig. She explained many others had turned down the opportunity to moderate the meeting, but she offered to take on the challenge, because, "I know the people of Eden Prairie are good people and the right thing will happen." (After tonight, her place in heaven was guaranteed.)

A representative of the Upgrala Hunting Club spoke first. "The Upgrala property consists of farmlands, grasslands, woodlands, and a wetland complex consisting of Upper and Lower Grass Lakes. It's one of the finest river and wetland complexes in the nation. The huge loss of our wetlands is a national disgrace. We have a chance here to prevent yet another loss by saying "no" to the proposed expansion. Clearly, it would be ludicrous and probably an illegal impossibility to locate a new landfill on the site.

"Recent water sample tests have found the natural seepage of river bottoms are contaminated. The only wildlife that would prosper from such contamination are rats, whose population near our homestead has burgeoned over the last decade.

"What you must decide is if the settlement compromise justifies the legacy of a further defiled environment."

A pro-settlement man spoke next. BFI complained that people who supported the expansion were too intimidated to voice their opinion at the last hearing. They insisted their position be given equal time, so a Mr. Johnson told the crowd, "Everything I have seen to date has indicated that our community has the deck stacked against us." (BFI's Irv had traveled around town and the Twin Cities telling any willing listener the expansion is a sure thing.) He advocated accepting the settlement package. The crowd booed him off the floor.

Another pro-expansion opinion came from Tom Vandervort, BFI's public relations man. (Oh well, if you can't find willing citizens, company executives will have to make do.) Tom gave a Bloomington address for himself, then complained, "There hadn't been much of a chance for people in this community to ask questions and get answers in a reasonable and forthright manner."

"That's because you're taking up all the time," shouted someone from the crowd.

He continued by telling us, "If you look at BFI and Woodlake objectively and carefully, you'll find they're good companies operated with integrity."

Tom received a resounding boo for that remark. When he finished a few sentences later, he got booed again. He wasn't the crowd's favorite spokesperson, but I was told BFI paid him handsomely for his aggravation.

Teresa Mathews, a mother and teacher, next recited the Boy Scout pledge, which she thought everyone should abide by, especially the powers-that-be in the room: "To do my best, to be clean in my outdoor manners, to treat the outdoors as a heritage to be improved for our greater enjoyment, and to keep my trash and garbage out of America's water's, fields, woods and roadways."

BFI's "Baseball" flew in from Chicago and was angry. He said scientific studies didn't indicate that the landfill received a large amount of hazardous waste in the past. "That's a lie!" he declared.

When Dr. Roy finished his short speech, he received the same treatment as his colleagues. An unidentified speaker recommended he, "Go back to Chicago!"

The next man commented, "To reopen the landfill would, in my opinion, be foolhardy at the least and unconscionable at worse, given its present conditions."

My good friend Chris of BFI then told us not to worry about health and safety problems at the landfill, since they had them fixed now. The people disagreed and booed him, too. Such is the price of defending a toxic dump. I knew he could stand a little rejection if he had my friendship and a fat paycheck every month. (The latter probably weighed in more heavily.)

I succeeded Chris. When I looked up from the podium at walls of people cheering me, my legs shook with emotion. For a brief few seconds, I smiled and bathed in the luxury of the applause. Then I directed my opening remarks to the council. I reviewed all the work we had done together, the path we had walked together to shut the dump and get the case in court. "So why the eleventh hour hesitation? Greater feats have been accomplished by the people. Who would have believed two months ago the Berlin Wall would have come crashing down?" (I had no idea Jerri used the same line.)

"It wasn't a king. It wasn't a president, and it wasn't an army that broke it. It was the people who refused to buckle under the injustices of the government."

I described my visit with a Polish Solidarity leader, and how I expected him to be some kind of "superman," and found instead "he was like all of us who care about the world our children will someday inherit."

I explained how the Poles accomplished their feat through a massive grassroots campaign that swept across the country, involving people from many different professions. "This same story is being played out in Eden Prairie," I continued.

"We have gotten the truth, we have gotten the people, and we are going to win.

"There are no walls high enough, governments powerful enough, or companies rich enough, to crush the spirit of the people who feel violated by injustice."

Turning to Eden Prairie officials, I said, "You, the City Council of Eden Prairie, have the opportunity now to be heroes and heroines of tomorrow in this precedent-setting case for environmental justice. But first, get back on track with the people. They have a voice through you, and they are running to victory."

The people rose to their feet and thundered their applause. I quietly walked back to my chair with my head down, thinking the people hadn't forgotten me after all.

The man who addressed the council next sympathized with city officials for having to put up with attacks from the people. He demanded to know if all the placards were made from recycled paper. He supported the settlement proposal, but no one supported him.

Then came Tom Brandabur, emcee of a presentation I labeled, "Truth or Consequences." He threw a bag of garbage on the floor, saying he felt more comfortable with this prop. "You know the BFI gentleman that spoke and gave his address?" he asked, "unfortunately sir, I'm a resident of the same office building you mentioned as your home address, which is about half way to the dump."

Barb Bohn walked up with a group of Girl Scouts who broke into a song sung to the tune of, "I'm Bringing Home A Baby Bumble Bee."

"Oh, my, I'm drowning in garbage from BFI, from BFI, from BFI. I'm being drowned in garbage and I don't like it one bit. What do you know, it's up to my toes. Yuck, oh gee, it's up to my knees. Ish, oh my, it's up to my thighs. Gross, oh fiddle, it's up to my middle. Yeck, oh heck, it's up to my neck. P.U. oh dread, it'll be over my head if we don't stop the dump. Say no to BFI."

"They just really wanted to speak tonight," said Barb, sweetly. She then continued alone and changed her tone. "BFI is, in fact' raping the citizens of Eden Prairie. The dump expansion rapes our quality of life. The dump expansion rapes us of a safe and healthy environment in which to live and raise our children."

Marlene Morgel, a homeowner living on the seventeenth fairway of the posh Olympic Hills Golf Course, suggested we change the name of Eden Prairie to Eden Dump. She threw $60.00 on the floor in front of the city council, the price of increased taxes to continue to fight the case, and shouted, "I'd rather fight and lose than yield to intimidation."

"I think we should throw our settlement in their garbage because it is...on second thought, I don't think we should throw it in the garbage, because it's hazardous waste itself," said John Habermaier.

Identifying himself as a former couch potato, Chip Smith said, "I'm stepping forward and carrying my part."

Tom Larson of the Minnesota Valley National Wildlife Refuge took issue with a fact BFI put out which says, "To date, there has been no reported impact on wildlife as the result of the operation of the landfill. Since no monitoring has been done, it's presumptuous to state there has been 'no impact on wildlife."

"I am Dr. Dorris Brooker," bellowed the next speaker. "I'm on the faculty of the University of Minnesota. I'm in the medical school. I'm a specialist. I'm a pathologist; that's the study of disease in maternal and child health. I'm concerned this evening."

Raising her voice even louder, she continued, "Doctor Roy, wherever you are from, this is Doctor Brooker from the University of Minnesota talking to you. It's my professional opinion the Flying Cloud Sanitary Landfill should be closed and stay closed, immediately and henceforth. It has benzene and 1,1,2,2, tetrachloride. These are known carcinogens that cause cancer, which causes birth defects. They are in groundwater. These are some of the most dangerous of organic volatile compounds."

Our citizen expert made a dynamic impression on the group. Today we still laugh about her public confrontation with BFI's "Baseball."

Fred LeGrand stated he was angry because, "I think this Homeowner's Association is affecting my property value."

The President of Knutson Mortgage took another view of property values. He consulted with an environmental engineering firm on how an expanded dump will affect his company. Because "BFI has defined the impact zone, it will be very difficult to make a loan in those areas. The mortgages will not be eligible for delivery to Ginny-Mae if the effect of the environmental hazard on the value of the property cannot be determined. They set the standards. They do not want mortgages that cannot be assured there will not be contaminated property for the 30 years of the mortgage."

Kevin Perry reminded everyone, "Right now you have a 30 percent chance of being able to close the dump if you do contest the case. But I assure you, if you don't contest the case, you have a zero percent chance of closing the dump."

In all, 25 speakers and 10 written statements opposed the dump. With only six people supporting expansion, three of them BFI officials, it looked like the anti-settlement group prevailed. I couldn't imagine the council voting against the people now.

Tonight's experience resembled the biblical story of the mustard seed, I thought. The seeds a few of us sowed years ago fell on good ground, grew, and bore fruit. The harvest was bountiful, and I felt privileged to dine at the banquet.

The mayor, however, wasn't happy with the state of affairs in his city. He wrote in the *Eden Prairie News*, "I protest the process. I protest the success of expedience and innuendo. I protest the new politics of slogans over substance. I will not capitulate to the new politics. Conscience prevents me from pretending to support and encourage those whose rhetoric has replaced reason. I protest."

Dr. Jean Harris, City Councilwoman, had a different opinion: "I thought it was just the best example of grass roots democracy I've seen."

Life moved on fast forward during my short stay in Minnesota. I attended a chamber of commerce breakfast that Dick Coller addressed, and before it was over, Jerri and I sped downtown to speak with Mark Andrews. Then Grant asked me to join him, Ric, and Linda to meet with Dr. Vincent Garry, a toxicologist, who the city hoped to hire as a witness if the MPCA granted the four additional issues.

After I explained the case to him in his small, crowded office piled high with papers and books, he just laughed. "You're kidding," he said. Linda assured him that, unfortunately, this wasn't a joke. She went on to explain all the technical aspects of the geology and hydrology. He couldn't believe the stupidity of expanding a landfill in a residential neighborhood. I pointed out there are more citizens living near this dump than any other one in Minnesota. Dr. Garry said historically it has never worked well to have garbage dumps near population centers; this is a public health issue. He agreed to testify if we needed him. I sensed he was the maverick type, not afraid to take a strong stand in court.

Dr. Garry observed that even two thousand years ago, the Romans disposed of their garbage two miles outside the city's gates. "When humans lost their solid waste sense in the 1350's, the Black Plague killed half the population of Europe," he said. "During that time, people threw their trash in the streets, providing local rats with generous food rations. Disease carrying fleas then found cozy quarters on the multitude of rats, and eventually jumped on people, infecting them." How ignorant, I thought, not to learn from mistakes of the past.

Grant also called the governor and convinced him to come out to Eden Prairie for a landfill tour. Although I was scheduled to leave before Perpich's visit, Dick asked me to postpone my departure and give the tour. I telephoned Ivan and asked him if I could extend my stay to do the honors. He agreed, and my father volunteered to watch my boys a few days longer. Also, Maureen Wilson gave me a skirt, because I ran out of clothes.

With only two days notice to prepare for the governor's dump tour, Homeward Hills resembled an ant hill in full activity. A group of mothers went to school and inserted flyers into the children's folders announcing the visit, then sent the message out through the telephone tree. The Environmental Committee determined the itinerary, and we each vigorously plunged into our assigned tasks to meet the deadline.

A fireside chat at the Collers with Governor Perpich was scheduled for after the tour. Taking a quick look around herself, Jerri declared her house a disaster area. After spending months preparing for the hearings, she neglected her homemaker duties. Hurriedly, Jerri called her neighbor Jeannie Thompson and asked to hire her maid for a day. "I'll take care of it," Jeanne said and hung up. Other mothers volunteered to bake bars and cookies for refreshments.

I left a message on Father Tim's answering machine asking him to welcome the governor at Pax Christi Church, since Rudy Perpich is Catholic. After my call, someone informed me Father Tim was away on a week's retreat and wouldn't be back on time. Oh well, I tried.

The day before the governor's visit, I walked over to Jerri's house to search some scientific reports for facts I could interject in the tour. When I knocked at the Coller door, a voice from inside the house shouted, "Come on in!" I entered, and there was Jerri, standing in her underwear on the upstairs balcony. She hadn't even had time to dress, and here I was barging in for information. I asked her where I could find the papers I needed. Jerri directed me to boxes of files where I began my search. The paperwork had multiplied many times since I turned over my files to Scott. There was a formidable job ahead of me locating what I needed.

Barely had I begun, than the doorbell rang again. Jerri bounded down the stairs to answer it, dressed this time. She opened it and found her neighbors, Jeannie Thompson and Donna Benz standing there with scarves tied around their hair. Holding buckets of cleaning utensils, sprays, and disinfectants, they sang, "Good morning, Merry Maids." Jerry groaned, "Oh my God, I don't want my neighbors to see my dirt."

In they came, and to work they went. The doorbell continued ringing all morning as more mothers came to offer help. With all this commotion, I couldn't do my research. I pleaded with Jerri to find me a quiet spot, so she directed me upstairs to her bedroom. I grabbed a box of papers and closed the door for a few hours to finish my task. When I emerged, the house was calm and immaculate, except the dining room.

In short order, clean dishes were unloaded from the dishwasher and dirty ones put in, bathrooms were scrubbed cleaned, throw rugs were taken away and laundered, carpet and floors vacuumed and washed, fresh poinsettias decorated the living room, and a pyramid of china cups and saucers was arranged on the coffee table next to plates of homemade bars and cookies. However, Jerri decided to leave the dining room untouched where dump papers were scattered across the table, on chairs, and in boxes over the floor. The Governor of Minnesota must see HHHA Command Headquarters unvarnished, so he could appreciate the enormous effort exerted by citizens to shut down the dump permanently.

That evening, I typed the tour on Jerri's computer. I was dead tired; my fingers would hardly move across the keys. It seemed to take me hours to complete the two pages. We ran off several copies on the printer to give to drivers of cars for their itinerary. I finally left the Collers late that night to get a few hours sleep before the 9:00 a.m. tour.

At 8:30 next morning, organizers arrived at Pax Christi Church. Dick Coller spread out maps and gave Joe Mengel, Grant, Slawek Michalowski our river bluffs expert, and me, orders on who should do what, when, and where. After receiving my assignment, I joined Jerri and Lynn Forster (the person who gathered 1000 signatures on a petition), at the church entrance to keep vigil for Minnesota's leading politician.

About ten minutes before nine o'clock, a panting Father Tim came barreling through the doorway. He mumbled something about getting my message and leaving the retreat early to welcome the governor. I smiled. Father Tim couldn't resist the opportunity to participate in the action.

Father Tim, Jerri, and Lynn Forster then stood together at the church entrance while I retreated to the conference room.

Shortly after 9:00 a.m., a black, chauffeured Lincoln pulled up to Pax Christi with the Governor of Minnesota and his aide. As orchestrated, Father Tim and the two women welcomed him. Then Jerri walked him back through the hallway and remarked, "Governor, where is your hat?" (It was bitter cold out.) He laughed.

A tall, broad, dark-haired, somber man of about 55 years old, dressed in a black leather jacket and dark trousers entered the conference room. He introduced himself as Governor Rudy Perpich. After we each presented ourselves to him, we reviewed BFI's dump on our maps and charts. After a scant, 15-minute discussion, we left in cars for the actual site. Jerri and I, plus Governor Perpich's aide sat in the spacious rear of the Lincoln, and Governor Rudy Perpich and his chauffeur occupied the front. Jerri and I giggled with excitement as we wallowed in one of the few perks of our job.

The rest of the group followed in several nondescript vehicles. Pitiful peons, having to travel in such impoverished style, I sighed. As we drove west on Pioneer Trail to Flying Cloud Drive, I began a monologue of the dump's history. Interrupting my speech was the first stop, the overlook located just past the BFI gate and adjacent to Schmidt's trailer. We piled out of our respective vehicles into the cold, and walked a short distance to a stone wall that stood on the river bluff. As usual the weather cooperated fully.

"This is one of the most beautiful spots in Minnesota," exclaimed the governor, as he scanned the valley with his eyes.

I took a place next to him and softly stated, "This is the largest urban wildlife refuge in the country. It's a stop off point for many migratory birds, and home to 400 heron nests. The rare harvest mouse inhabits this fragile ecosystem, too," I said. "Contamination has been detected in seeps that bubble up into the refuge." After my several minute discourse, I relinquished my place to our river bluffs expert. In accented English, Slawek Michalowski, who, like Governor Perpich, is Polish, discussed the unstable nature of the slope, and pointed out a crooked tree which he said indicated the bluff was slowly creeping into the river. Joe explained the geology of the area, then we quickly returned to our warm cars.

The Governor's chauffeur retraced our route on Flying Cloud Drive past the airport, while I explained the dangers of a dump and an airport co-existing closer than FAA regulations permitted. We then turned right on County Road 1, and drove to Homeward Hills Road where we took another right and descended to the park. Jerri and I talked about the population density of the neighborhood, quoting from an EPA report which targets a four-mile radius around a dump as a risk area. Governor Perpich said nearly nothing and just listened. As we approached the park, there must have been a thousand people standing outside the warming house waiting for him. Others came running and driving their cars across the frozen prairie to hear the governor speak. He was impressed. "Why the hell are we dumping garbage here? Isn't there someplace else we can put it?" he questioned.

Before the aide could reply, Jerri reached up and squeezed the governor's shoulder and said, "We wondered that for so long." I thought my ears would fall off with Governor Perpich's remark.

Our car pulled into the parking lot, and he just leaned back in his seat and said, "There are so many people. I have nothing prepared."

When Governor Perpich got out of his car he exclaimed, "Look at all the children!" They were running in all directions, delighted with the commotion. I recognized many of them and several cried out my name. I happily waved and said, "Hello."

"These are my friends," I cheerfully explained to the governor. He was visibly moved by the people surrounding him. I walked slowly with him to the platform, while Jerri ran ahead with Dick to make sure the sound equipment was in order.

Before Governor Rudy Perpich addressed the crowd, Jerri welcomed everyone to the rally like a polished politician. Jerri was warmly dressed in her ski jacket and hat. While listening to his speech, I stood solemnly alongside the governor in my red wool coat, hatless, following Rudy's vain example. How impressed I was with my place! How long I had waited for this moment!

Governor Perpich made only a few comments, then descended from the makeshift stage and waded into the crowd like a populist. I accompanied him. At that moment, my good friend Sharon Papic came running over and gave me a hug. "It made me teary eyed to see you last night," she said. "It felt so good to be appreciated."

I resumed my place next to the governor and put my hand under his arm to guide him through the crowds. To this day, I'm embarrassed by my boldness, but he appeared dazed and seemed to need direction. He paused to listen to Steve Frick explain the toxicity of the water contaminants, and another woman discuss her child's asthma. His eyes darkened. The residents knew the facts well, and lobbied their governor hard. He never smiled.

Governor Perpich had an appointment after the tour, so I gently whispered in his ear that we had other things to show him before he left. I ushered him to his car and he obediently came. Our motorized caravan left the park, traveled up Silverwood to Winter Place, then turned right to

Sandy Point, the street closest to the dump. Here we showed the governor the monitoring wells in undeveloped lots next to the landfill, and pointed out the limit of the underground plume. Two lots away stood a home. We explained that no one wanted to buy the undeveloped land, because of its proximity to the dump. Perpich was silent.

Dick periodically jumped out of his car and ran up to us to check on our progress. Since we were ahead of schedule, Jerri decided to take a little detour to the bluff area where a sewer pipe fell into the valley during the 1987, "storm of the century."

After a good look at the collapsed slope, we headed for the Coller's home. A roaring fire in the living room took the chill off our bodies as we settled down to chat with the governor. The press waited in cars outside, so we could speak privately.

Hot coffee, bars, muffins, and fortune cookies were passed around. The governor took nothing but coffee, so I offered him a cookie. He cracked it open, read his fortune and laughed heartily for the first time all day. It said, "Confucius say: 'Close the dump and you will have good fortune.' "

After we thawed out, Dick led the conversation. He got up on his knees facing the governor across the coffee table, and explained that people felt helpless trying to defend their neighborhood and their children's health against the forces of big business. Worse yet, the government endorsed BFI's plan to expand the dump, explained Dick, very sincerely. Governor Perpich thrived on protecting the Davids of the world from Goliaths, and Dick played to this need.

I described yellowed and sickly people I had met at the CCHW convention, who lived near toxic dumps. I then appealed to the governor to stop the dump expansion madness in Eden Prairie before it reached the same ugly impact on the lives of our people. He listened to attentively to my comments and those of others.

When everyone finished speaking, Perpich told us the MPCA Board won't necessarily follow its staff's recommendation to expand the dump. "The board members are very independent-minded people," he said.

Jerri invited the press in, and the governor stated for the evening news, "The MPCA board always gives the benefit of the doubt to the environment. It's grossly unfair to assume they're going to rubber-stamp anything." Since the governor appoints the people to the board, he subtly

267

nudged them in our direction.

As Rudy prepared to leave, he whispered in my ear, "Good job!" I was pleased to earn his respect, and was deeply impressed with him as well. This man really did care. After the governor left, we glowed around the fire and declared our day a complete success.

The vote

On December 12, 1989, Dick Coller contracted diarrhea, and his wife Jerri broke out in hives. On this day, the Eden Prairie City Council was voting to accept or reject the settlement. Dick and Jerri nervously commiserated about the turnout for this last hearing. Naturally, there was a major community event conflicting with the meeting. And just maybe citizens would stay home because they were fed up attending meetings.

That afternoon, Jerri received a call from Council Member Jean Harris. Jean had labeled herself the swing vote and hadn't yet taken a position for or against the settlement until now. She telephoned to say, "Don't worry, thumbs up. Go get 'em!"

Jerri, dressed in her black dump skirt and white polka dot blouse, went early to the meeting with Dick to set a little box containing two fortune cookies at each council member's place. "We left no stone upturned," Jerri said. Then came the usual wait for people to fill up the room.

A crowd of 400 people packed Wooddale Church. At 7:30, the mayor brought the meeting to order and explained the proceedings. Dick dispensed with his chanting and cheerleading activities for the night. He said BFI, the citizens, and the community at large would each be given a half hour for case presentation. BFI representatives spoke first.

Chris Deitzen began by discussing the dump's history, just in case someone had newly arrived in town unaware it dominated newspaper headlines for eight years. Then Dr. Roy stated "facts" on how we'd lose the contested case hearing. Proof in point, "Will Woodlake comply with the applicable statues? We will," said Roy assuredly. No supporting information was given.

Dick Coller proceeded to the microphone next to address the city council. "You said we must turn out the people, and not just 40 from our neighborhood. We turned out thousands. You said the second meeting is more important, because the first one was in your neighborhood. We again turned out thousands. You asked us to speak here again tonight. My God, you'd think a person would be convinced by now!"

Dick implored the city council to believe the testimony of a University of Minnesota doctor, and a local firefighter who put his life in danger extinguishing a landfill fire, rather than believe BFI's Dr. Roy of Chicago. Dick continued for several minutes, reiterating remarks made by qualified members of the community, then discussed the money issue. "If the city spends 2 or 3 million dollars fighting the contested case hearing and loses, they still stand to get $18 million from landfill fees if the expansion goes through.

"The argument of spending money is transparent. We gain nothing from the proposed settlement. We lose nothing by continuing to fight. By giving into pressure tactics and intimidation, we lose our self-respect and honor, and most of all, our chance to close this threat to the health of our children forever. Please close the dump."

Grant Merritt succeeded Dick and predicted the city and its citizens would win the case, because of "a grand outpouring of feelings against the dump idea of BFI." He then presented a letter from the federal Environmental Protection Agency which stated: "Under the proposed criteria, expansions of existing facilities would be treated in the same manner as new facilities."

BFI's hope to expand the landfill now diminished considerably, because its new facility would violate proposed federal regulations. However, because the regulations weren't yet in effect, there was always the possibility BFI could pass its proposal before the proposed federal siting criteria became law.

Jerri spoke last, but not least. She said, "They (BFI and its Woodlake subsidiary) may be a big corporation from Texas; they may have a lot of money, but when they took on Eden Prairie, they bit off more than they could chew." She ended her citizen testimony with an appeal to the council to go forward united with the people. She cited an Aesop's fable: "Separate we can be easily broken, together we are strong and cannot break."

Steve Keefe, Chairman of the Metropolitan Council, left his seat beside his BFI cronies to speak first for the community at large. Of course, he emphatically stated, "If this landfill expansion isn't approved by the MPCA, landfill capacity in the Twin Cities will run out sometime in 1993, which isn't very far away." That argument impressed no one. Did Steve really believe the good people of Eden Prairie would voluntarily offer up their community as a dump site for metropolitan garbage?

Even less impressive was his comment that his "staff and some council members have some reservations about the settlement, because they think it's too generous, actually."

The meeting transcript reads, "Adverse response from the audience."

Jackie Hunt of Greenpeace spoke next, supporting the citizens. Tom Bierman, Soccer Commissioner and my sons' Pierre's and Neil's former coach, said he assumed things were going along the right path, that city fathers had taken a strong position against BFI 's expansion, especially when an arrogant letter from BFI arrived. Tom Schwartz, formerly an employee of Allied Chemical, stated he was certain chemicals from his company made it into the dump. Donald Berne remarked he had been around the area for the last twenty years and had seen a lot of stuff go into the dump; "There's everything under the sun. It's nothing but a poison trap."

That was it; no more testimony.

"Your honor, said Councilman Anderson, I'd like to take this time to make a motion. I move the following resolution to reject the proposal by BFI: Resolve that the City of Eden Prairie, Minnesota, hereby denies approval of the settlement proposal for the contested case hearing expansion of the Flying Cloud Landfill outlined in the certain documents entitled Woodlake Sanitary Service Proposal dated October 17, 1989."

Jean Harris seconded the motion, "Because the technological barriers and assurances which BFI have offered to us go a long way to abate some of the concerns I have had, but not all of them." She explained her six-month old car is already leaking oil on her garage floor; "none of us can guarantee that technology will protect any of us over the next hundred or two hundred years."

Patricia Madame fell in line with her "nay" vote, because she also didn't trust the engineering systems to function perfectly, and reminded everyone that a defective ring in the Space Shuttle caused it to blow up.

Doug Tenpas had already publically opposed the settlement, so his vote was no revelation.

Last to speak was the mayor. First, he thanked everyone for their participation, then scolded Doug Tenpas for not thanking him for going to the settlement talks. Councilman Tenpas apologized for this omission. Mr. Mayor next rambled on about how difficult the decision was, how there were many sides to the case, how he didn't think the city could win, and how the MPCA would probably issue the permit anyway.

Then the vote was cast. It was 4 to 1 opposing the settlement. The mayor cast the only pro-expansion vote. The crowd was speechless.

Dick Anderson finished up by speaking movingly about his long, proud residence in Eden Prairie, where he had raised his children. He said he wasn't convinced there wouldn't be health problems in the children living near the dump if the expansion were granted. Also, he was concerned about damage to the wildlife refuge. "You gave me your opinion loud and strong; you made it easy for me to cast my vote for rejection of the landfill."

While everyone hugged and kissed, a grim mayor left the room under tight police guard. Jerri cried and thanked Dick Anderson for his resolution. Grant and the citizen leaders moved on to Applebees Restaurant for a celebration. Grant kindly telephoned me the news; I had worried the whole day about the outcome. Many friends lined up to speak with me, and I appreciated the opportunity to hear the excitement and relief in their voices as they related the evening's events and their shock at the mayor's vote.

I believed it was just a matter of time now until we won the case. We just had to find the right angle or hook with which to finish it off. But what would it be?

13 | TEARS AND CHAMPAGNE

"**I** attended a little luncheon with the admiral," related Sidney, when she told me about her meeting with the chief administrator of the Federal Aviation Administration (FAA). "Rudy (Minnesota's governor) was there too, so I talked to him and he said, 'Let's sit together, Sid. Stick with me.' I did and planted myself in the chair opposite Admiral Busey, and gave him the whole landfill story. I told him the distance between the airport and the landfill, and he picked up on the fact that the landfill was closer than regulations permitted. He was concerned, and asked me to get it all in writing. Before the day was out, I had a letter off to him."

In 1982, BFI received a variance from FAA laws, and on December 14, 1989, Sidney requested that the agency reverse its position. We waited impatiently for the response.

February 28, 1990, a lengthy letter arrived from the Admiral explaining again the regulations we all could recite from memory, and saying the MPCA must seek FAA approval for the landfill expansion. "If such approval were sought, we would withhold it on the basis of

273

incompatible land usage," said Admiral Busey.

"Clearly, we do oppose expansion of this landfill, but the FAA does not have the authority to close landfills, and is not responsible for ensuring that the MPCA adheres to its requirements."

Sidney telephoned me and gleefully told me the good news. "This could be it," I declared enthusiastically.

"I think so," Sidney said softly. "It could be the final nail in the coffin."

City lawyers asked Judge Klein for a summary disposition to decide the contested case in our favor without hearing further arguments. The reason: the expansion violates FAA and MPCA rules regarding distances between airports and landfills. Lanny, the permit supervisor for the MPCA, told the press that his agency considered the 1982 FAA approval for the expansion adequate, and they didn't plan to seek further FAA endorsement. The president of BFI said he was willing to study the habits of birds around the landfill and plant a series of "bird wires" on poles to divert them from the area. Jerri Coller said, "It's another example of their incredible gall."

With heated remarks flying across the pages of the newspapers, Judge Klein took time to respond to the summary disposition.

Meanwhile, other carpenters hammered nails into the other end of the BFI coffin.

Grant held a meeting of the newly hired scientific experts in his office. Dr. Vincent Garry, Toxicologist; Dr. Joe Mengel, Geologist; George Noble, Civil Engineer; Linda Lehman, Hydrologist; and Kurt Heikela, Chemist, sat down to discuss their strategy for the remainder of the contested case hearing. I imagine the average IQ of the group was about 140; I regretted my absence at this "meeting of great minds."

They chose a two-pronged attack: site suitability and health effects. In short, is this site suitable from a public health standpoint?

To determine the answer, the scientists requested additional groundwater studies. Understandably, they didn't trust BFI's test results done at Pace Labs, so they contracted with another laboratory to do the analyses. Our scientists also wanted to decide where to take the samples. We had the distinct impression BFI tested water where they expected to find the lowest levels of contaminants. The lawyers appealed to the city for funding, and to Klein for a postponement in the case. The city budgeted $100,000 for the testing, and the Judge agreed to delay the hearing.

At this time, an idea popped into the head of Councilwoman Patricia Madame, too. If we're serious about winning the case, she reasoned, we had to get the town's businesses involved. Why not set up a committee of influential business leaders to advise the community on the landfill expansion issue? Patricia nabbed her good friend, Sidney Pauly, and together these women hosted a "little breakfast" to enlist the help of Eden Prairie's CEO's and company presidents. It worked.

Soon after, The Landfill Advisory Board was established. It consisted of Glenn Wilson, President of Knutson Mortgage Company; George Butzow, Chairperson of the Board of MTS Systems Corporation; Robert Cox, President of Rosemount, Inc.; Al Lyng, Business Banker of First Bank of Eden Prairie; Al Lange, Vice President and General Manager of Elliot Flying Service; The Honorable Sidney Pauly, Minnesota House of Representatives; Council Member Patricia "Madame" Pidcock; Council Member Douglas Tenpas; HHHA leader, Jerri Coller; and interested HHHA members. The dynamism of this group charged up the community on the dump issue.

Patricia Madame excitedly telephoned me in March with her brilliant plan.

"So what are you going to do with all these people?" I asked.

"We're going to disseminate information to show we have a basis of fact for our position opposing the expansion."

"Great!" I said.

And they did. They retained the services of Lobbyist Pat Forciea, and public relations person, Kathy Kennedy, to pursue their mission. The first thing the group accomplished was putting together a video of the landfill story, using experts and citizens to state the facts, using the dump as a backdrop. It begins with a close-up of someone's hand sifting sugar sands on the bluff to illustrate substandard soils under the landfill, and ends with neighborhood children playing in a sandbox next to the dump's barbed wire fence, demonstrating the dump's proximity to the community. As the scene fades, the narrator warns, "Let's not bury our heads in the sand or our garbage in it, either. These guys have to trust our judgment."

Everyone spoke with enormous concern on the tape. There was a shot of neighbors huddled in Coller's living room, brainstorming at one of their many strategy sessions. Dr. Garry didn't laugh this time when he said, "We learned from history that it is not good to have trash by people. Yes, the engineers will tell you everything is fine, but we biologists don't agree. Biology is an inexact science." This collaboration between scientists, business leaders, a public relations firm, and the citizens didn't produce a Hollywood blockbuster, but, it clearly stated why the Flying Cloud Superfund site shouldn't be expanded closer to homes and a wildlife refuge. Every Minnesota senator and representative was targeted to receive a copy of the tape, as well as influential business people, environmental groups, investigative reporters (I-Team), and others.

In addition to the video, the public relations firm set up a letter-writing campaign to keep state and local newspapers filled with disgruntled Eden Prairie folk. They also arranged for HHHA leaders to appear on a TV talk show and on cable television, and be interviewed on radio, too. Jerri gladly relinquished her duty of contacting the press.

With more free time, Jerri sought to expand the group's political support. She arranged for meetings with U.S. Representative Bill Frenzel, U.S Senator Dave Durenburger, and gave Frenzel a dump tour. Jerri remained convinced that if political leaders could see the dump's location, they'd understand why citizens were so upset about the expansion.

On the second anniversary of the dump closure, April 4, 1990, Jerri staged a candlelight vigil at the park. Dale, Pax Christi's guitarist and singer, sang for a group of 75 people as the sun set on the open prairie. Councilman Doug Tenpas read a poem. Jerri said, "It was good to take a few quiet moments to be thankful the landfill was closed, and ask for continued support."

A full page profile of Jerri's daughter appeared on the front page of *The Sailor* newspaper. In the close-up, Renee, 6, angelically is gazing into her burning candle, held carefully in her tiny hand. Her long, disheveled, blonde hair falls in wisps across her cheek. Another little girl next to her and slightly out of focus, is staring at her candle, too. The picture is a study in contrasts. It's startling, yet soft. The haunting innocence of the children, each holding a single candle in the night, could prompt one to think of endless symbols.

Without knowing what photo the newspapers would print for the first page, I urged our citizens forward with these words, "When you become fed up with endless meetings, letter writing, phone calling, and other continuing expenses, just look in the eyes of your children and remember it's for them you're sacrificing your time and money, and they're worth it. So press on. Victory is at hand."

The twentieth anniversary of Earth Day, April 22, 1990, gave HHHA another opportunity to celebrate and make the newspapers. For the weekend, Jerri organized a tree-planting, a compost demonstration, and a concert to raise funds for the neighborhood environmental cause.

BFI's last stand

When I arrived home late from my graduate class one Tuesday in March, Pierre tore open the garage door to tell me Sidney Pauly just telephoned. "She's very upset," he said, visibly agitated. "BFI hired 22 lobbyists to push through a bill."

Oh, my God, I thought. BFI must really be desperate. How could we ever beat numbers like that?

I telephoned Sidney, but couldn't get through to her. I figured she must be out lobbying around the clock. As I learned later, BFI tried inconspicuously at the last minute to slip an amendment into the Waste Management Bill, granting the expansion of a landfill over a local government's objection, if it's approved by the Metropolitan Council, the Minnesota Pollution Control Agency, and the county. This legislation was obviously aimed at Flying Cloud. With the city and its citizens aligned again, BFI had few options left with which to expand its operation. This was a desperation move.

At the request of Sidney, the city council hired Ron Jerich, "a democrat with good connections," to defeat the amendment. "Ron had friends where we needed them," said Sidney. So Ron, Sidney, Senator Don Storm, Pat Forciea, and Jerri Coller and friends hit the halls of the capitol to defeat the amendment. "We didn't trust the phones; we just tiptoed around," Sidney told me later.

A *Star Tribune* reporter cornered Sidney at the state capitol to get her reaction to BFI's latest legislative push, and our normally mild-mannered Sidney burned the reporter's ears with her remarks. She accused BFI of heavy-handed tactics. "It's the kind of stuff they're famous for doing; they act like a bunch of thugs."

277

In her family van, Jerri drove five neighborhood mothers dressed in Easter suits, to the state capitol to influence decision- makers. They called lawmakers out of session one by one to give them the lowdown on the dump (pun intended). With all these well intentioned women outfitted in their Sunday best, how could these politicians resist their pleas for help? They couldn't.

BFI inserted its clever, narrowly defined amendment only in the Senate bill, where it believed it had a chance to win. BFI did win, but by only one vote. However, over in the House, Sidney and the Eden Prairie lobbyists had all their bases covered. Willard Munger, an ardent environmentalist and Chairman of the Environment and Natural Resources Committee, refused to give the bill a hearing in his committee. Sidney sent out a letter to her fellow House mates and urged them to defeat any attempt to push the issue on the floor.

BFI lobbyist, Jim Erickson, realized the bill was blocked by the House, so he pulled it. Losing in the legislature three years in a row would have been a strike out, so BFI preferred to quit the game instead, hoping to score the following year.

With the legislative battle over, and while lawyers wrangled with the potential for ending the case on the FAA issue, Jerri, Pat Forciea, Cathy Kennedy, and Sidney became information disseminators. They distributed dump literature and videos to key political and business leaders, and gave guided landfill tours.

During her public relations campaign, Jerri learned the governor was coming to town for a highway meeting. She arranged to be there also, to do a little lobbying on another city matter. After Jerri arrived at the meeting room, she scoped out the place and watched Perpich move around the room. "He had to exit by a certain ramp," she told me "so I positioned myself (all 5 feet of her) at the end of it and waited."

As the 6' 3" politician strode toward her, Jerri blocked his path. He put his arm around her shoulders and asked, "What can I do for you?"

"I asked him to put his opposition to the landfill expansion in writing. He agreed to do it." (Unfortunately, he never got around to it.)

After several months of deliberation on the FAA issue, Judge Klein rejected the city's demand for a quick end to the hearing. He concluded in legal language "The Summary Judgment Motion is not ripe at this time, and therefore should be denied, without prejudice to the filing of a later motion after circumstances have changed. There is a genuine fact issue regarding 'FAA approval.' "

City attorneys then met with lawyers representing the MPCA staff and board to discuss the problem, because the landfill also violated the agency's regulations. The MPCA lawyers asked Ric Rossow to sign a resolution supporting Klein's denial of the case. Ric refused.

During this period, I wrote to MPCA Commissioner Willet, informing him it was time to permanently shut the dump. An expansion was wrong, based on information I learned in my graduate classes, I said, and would set a dangerous precedent if granted. I concluded the letter by saying that as long as I existed on this good earth, I'd fight with all my resources to stop the Flying Cloud Landfill expansion. In return, I received a letter outlining current issues in the contested case hearing.

Sidney, Patricia Madame, and I brainstormed constantly by phone and came up with another avenue by which to use FAA's position. Since federal EPA regulations were also violated on this airport matter, we thought this agency might agree to write a letter to the MPCA requesting them to stop the permit process. The three of us thought Ric should call the EPA to see if this could be arranged. We were so close, and yet we couldn't get that last nail in the coffin.

The EPA agreed to review the case and consider the FAA letter, but warned the agency doesn't enforce these particular regulations. We all wondered that if the federal government, the state government, and the judge wouldn't enforce the laws, then who would?

While the EPA deliberated on the matter, city officials and Ric appeared before the state Metropolitan Airport Commission (MAC) to ask it to follow the lead of the Federal Aviation Agency (FAA) in opposing BFI's Flying Cloud Landfill expansion. (The FAA threatened to withhold funding promised to MAC if MAC didn't support FAA's position.) "We feel this is basically an MPCA issue," said MAC Commissioner Tim Lovaasen. Obviously, the buck didn't stop here either.

Finally, the city and its citizens asked the MPCA Board to address this issue at their meeting scheduled for June 26, 1990. This time the city conducted a big publicity campaign to get the people downtown.

Would this eight year nightmare ever end? Perhaps this might be it, I thought. The facts were there to support a total shut-down of the dump, and the board seemed sympathetic to the citizens.

One and a half years had elapsed since our family left Eden Prairie, and the boys wanted to run through the old neighborhoods again to see their friends. Therefore, a vacation to Disneyland lost out to a car trek back to Jack Pine Trail. With this meeting coming up, I decided to visit during the week the MPCA Board voted on the FAA issue. Also, a public hearing was scheduled in Eden Prairie several days before the board meeting, to take public testimony on the second set of contested case issues. We anticipated arriving in Eden Prairie the evening of this hearing. With only an hour to spare, we made it.

Ivan and the boys stayed behind at Brenda's house, while I drove to Hennepin Technical Center to present my term paper, "The Environmental Effects of Municipal Landfill Gases." As I walked into the familiar auditorium, I saw Judge Klein sitting at the front table and walked down the steps to say "hello." He smiled and seemed very pleased to see me, and asked how I was doing and what I was doing. I told him about my graduate classes and explained I'd testify this evening, using some of the information I learned from them.

After my few minutes exchange with Klein, I found Grant, Dick, and other HHHA leaders in the front rows, and asked if I could join them. Dick said there was no room for me, and pointed to a seat over on the side where I could sit. I declined his offer, since I wouldn't be able to see well from there, and walked back to sit with Roger and Sidney Pauly on the steps. They hugged me and welcomed me to join them, which I did.

The hearing began, and HHHA leaders testified first. Dick got up and gave a good technical speech on why current specifications for BFI's expansion were less than adequate to protect the environment and people. He handed Judge Klein the drawing of the triple-lined landfill I sent him from my Solid Waste Management class. Klein asked him where he got the information. Dick didn't answer. I whispered to Sidney that I had given it to him. The Judge again asked Dick the same question. Dick still said nothing.

Sidney then urged, "Raise your hand, Susan, and tell Klein." I did, and Dick threw up his hands in disgust. I was deeply hurt by his reaction, and Sidney seemed surprised also. I now hesitated to testify. Sidney encouraged me to speak anyway. However, I decided to wait until the TV cameras left before I spoke. I wanted to avoid giving the impression that I hungered for the limelight.

Finally, just before the break, I wildly waved my hand and Klein called on me. He smiled broadly as I approached the microphone and swore me in. I discussed the results of my graduate paper which I said contained a bibliography of 28 sources, "slight overkill for a master's paper." I stated that BFI's proposed expansion on this poor site "insults our intelligence." And BFI couldn't guarantee that its engineered systems would never fail, thus the population adjacent to the landfill was at risk. My talk received a tremendous round of applause, and, "The Environmental Effects of Municipal Landfill Gases" was entered into the case as exhibit 1011.

Following the hearing, many people came up to meet and thank me. Some of them I didn't even know. It was heartening.

MPCA's meeting followed a few days later. Grant assured us an MPCA lawyer gave the green light for citizen testimony. Four FAA officials flew in from Washington to speak, also.

Hours before our departure downtown, chaos reigned at the Collers. Their air conditioning broke, and it was sweltering inside and outside the house. Jerri called Betsy to drive her downtown so she could practice her speech in the cool quiet comfort of the car. I volunteered to ride the buses and help out with the people.

At Pax Christi, while waiting for the buses, I saw many people I had worked with over the years. There were some new faces, too. We all laughed at the prospect we might be doing this until we had grandchildren. Like cattle going to slaughter, we lined up, boarded two school buses, and left for St. Paul. We arrived ahead of Jerri, so I saved her a front row seat while I talked to Grant about strategy.

Three hundred people packed the board room. Many men, anticipating good news, took time off from their jobs. Expectations were very high, and the press was ready to record the moment.

An MPCA lawyer presented the issue before the board, then members of the board put their heads together in a long private conversation. What happened? We all looked at one another dumfounded. Wouldn't the people be given an opportunity to speak?

After a several minute powwow, the board chairperson stated the board would vote on the issue without public comments.

Unable to contain my rage, I stood up and shouted out, "Wait a minute, this is the United States of America, isn't it? These people were promised an opportunity to speak, and now they must be heard."

Jerri's face was bright red. Grant told her not to speak out; just be quiet. Jerri obeyed.

However, Bonnie Swaim stood bolt upright and yelled, "We want an opportunity to present our case." I was shocked. Was this the same woman who once told me she couldn't participate in the bus blockade; she'd stay home and pray instead? Yes, this was the very same person who got angry enough to throw fear in the toilet and demand justice.

Despite our protests, a vote was quickly taken. The board ruled against enforcing their own airport-landfill proximity regulations; they wanted a court of appeals to decide the constitutionality of the regulation.

There were many irate Eden Prairie folks in the MPCA board room that June 26, 1990. I was perhaps the most chagrined. I had traveled 1,200 miles to watch a fiasco. The crowd of citizens left very angry. Several board members followed us into the hallway to speak with us. Van Ellig, the young, good-looking lawyer who often supported us, told me he didn't appreciate my outburst. "What else could I do?" I said. "If people were promised they could speak, and then were denied that right, one has to shout out or be trampled." He quietly agreed with me.

"This whole thing makes a mockery of the system," Jerri told the press. Once again, the PCA has spit in the face of the people, and we're sick of it."

"Chairman Dr. Foley panicked when he saw the room full of protesters and representatives of the press, so he decided to avoid a confrontation by eliminating citizen testimony," I said to the *Sailor* newspaper. Jerri and I wrote scathing letters to the editor denouncing the behavior of the MPCA Board.

The Court of Appeals met July 16, 1990, to hear arguments on the Flying Cloud Landfill case. They expected to deliberate several months before they reached a decision. Again, no one had the guts to stop the buck.

The contested case resumed. One of the first witnesses to testify for the city was Dr. Clyde Hertzman, an epidemiologist who conducted a thorough health study on residents and workers living near or working at the Upper Ottawa Street Landfill in Hamilton, Ontario. The landfill is very similar to Flying Cloud. Chronic bronchitis, daily coughing, combined respiratory problems, narcotic symptoms, mood disorders, difficulty in breathing, skin rashes, and muscle weakness were detected in these two groups of people.

On learning this, the Eden Prairie City Council ordered a moratorium on development around BFI's landfill. The developer who put in the subdivision where I formerly lived, now was trying to build homes near the other end of the dump. The city squashed his plans for the moment.

BFI called its experts to court also. Rather than bore the reader with all the technical tedium of the case, I'll provide just a few "facts" given by two of BFI's scientists. The most impressive was George Tchobanoglous. He wrote the book, *Solid Waste*, the bible of the industry and the book undergraduate and graduate students like me use for studying landfill design. It's from this book I learned about basic site suitability. Dr. Tchobanoglous is a graduate of Stanford University, and regarded as a world expert in his field. When questioned about the suitability of the Flying Cloud site for an expansion, he stated, "After reviewing all the regulations, specifications, and data, it's my opinion the landfill site is suitable for the proposed expansion."

"What is the basis for your opinion?" questioned our lawyer.

"The information and data contained in the reports I've read, my own experience as a practicing civil and environmental engineer, my experience in gathering information I've used for the textbook I've written in the field, and additionally, my professional experience gained as a board member of the Integrated Waste Management Board of the State of California."

With an industry expert willing to make statements like this, I now understand why our Planet Earth hurls toward ecological disaster.

Brian Murphy, an engineer hired by BFI to make soil borings for the clean up system, presented a chart showing his placement of holes. A whole section of the slope near the crest was omitted. Strange, thought Grant and Joe. Why would he do this? They found out later.

Dump crusading began to exhaust Jerri and Dick, now. Dick worried his wife would have a nervous breakdown working such long hours everyday. On top of all this aggravation, Dick's mother died in May, after Jerri had flown out east to help care for her. Jerri was tired. Once when I spoke with her on the phone, she said she and Dick were looking at some land; they wanted to build another home somewhere else, away from the dump. She didn't know if she could put up with the fight much longer. I understood fully. I once nearly quit the crusade, just before the dump was ordered shut.

In early August, Dick took the family to the New Jersey shore for a vacation. No forwarding address or telephone number was given anyone; Jerri needed a rest.

On August 1, 1990, as lawyers and expert witnesses droned on in court, a crew of men drilled holes at BFI's landfill to install a barrier well clean up system. While drillers worked 100 feet south of the landfill boundary, trouble erupted on the crest of the bluff. Several small explosions and fires occurred at the rig site. The operation had to be shut down, because continued drilling jeopardized the worker's lives.

The workmen figured that methane gas generated by underlying garbage caused the explosions. For several months now, drillers had hit garbage along the bluff, in an area beyond the landfill perimeter. They noted this fact in their logs. On seeing these entries, the manager asked that the log entries be changed to "fill." Garbage outside the landfill boundary is a violation. The change was made.

The explosions were now reported to the newly hired landfill manager. (The former manager retired for health reasons). An internal investigation was begun. A BFI methane technician searched the files and discovered, hidden in a drawer, laboratory analyses dated November, 1988, showing methane gas had been detected at significant levels along the bluff where drillers had found garbage.

Incidentally, the trash was found in the area where Brian Murphy neglected to test the soil, and Bill Schmidt had compacted garbage to repair the 1987 bluff slide. Yes, Dwain Warner, you were right. Methane was seeping through the bluff.

A call to Pace Laboratory confirmed that testing was done in November, 1988, and the former manager had phoned to know the results immediately. When he was told gas levels were high, he instructed Pace not to send the information. The laboratory agreed. In March, 1989, Pace finally sent the analyses to BFI for payment. The manager shoved them in a drawer.

This news undermined the effectiveness of the entire clean up system. The just installed methane system couldn't extract gas generated outside the landfill perimeter, and the barrier well pumps couldn't be installed in garbage. Was this disclosure of withheld information and falsified laboratory analyses just the tip of the iceberg?

A very shaken and pale Chris Dietzen submitted the findings to Judge Klein in court. To his enormous credit, Chris made a very painful decision that cost him the case and might damage his reputation as one of Minnesota's top trial lawyers. I was told Chris is a deeply religious man and father of several children. Evidently, his conscience forced his hand. He pounded the last nail into BFI's coffin.

Klein told everyone in court not to divulge the information. He postponed the hearing indefinitely and granted the city the right to be present on the landfill property during tests, and to suggest where soil borings should be done.

September 5, 1990, the Minneapolis *Star Tribune* ran the story on its front page. Dean Rebuffoni revealed the suppression of evidence and falsified logs to the public.

BFI Houston headquarters shook again. This time the quake recorded off the Richter scale. Damage control experts swung into riot formation. Tom Vandervort was charged with presenting a sane, controlled sense of order to the outside world. He took his orders directly from Ruckelshaus now.

Meanwhile, the Coller family calmly returned from their vacation to find the neighborhood reeling with the fast paced events. Ric Rosow and Jerri quickly called a press conference to respond to BFI's withheld information debacle. Ric angrily demanded that the contested case be dismissed: "BFI isn't fit and able" to run the operation, he said. "If false or incomplete information is presented with a permit application, it's grounds for denying a permit."

Jerri said, "We're livid and we're outraged. We're tired of being ruled by cover-ups, lies, deceptions, and constant threats."

September 11, 1990, Jerri Coller, dressed in black shorts and a white blouse, set out early in the morning for Cub Grocery. She decided to stop by city hall first to pick up some landfill tapes. On her way into the building, she noticed two men leave in a red Bronco. After she walked through the door, she announced to the secretary, "I'm Jerri Coller, and I'm here to pick up six videos from Craig Dawson." The secretary gave her a vague look.

Then City Manager Carl Jullie, Assistant Craig Dawson, and Secretary Joyce Provo, asked Jerri to sit down on a chair they brought her. Jerri obliged, holding her grocery list. (What a dedicated homemaker!)

"They quit! BFI quit!" said Craig.

"That's not funny" replied Jerri.

"I'm not joking," Craig said.

"Didn't you just pass those men on your way in? They were from BFI," added Joyce.

They gave Jerri a letter to read. It said: "Please take notice that the Applicant, Woodlake Sanitary Service, Inc., (WSSI) hereby withdraws its application to expand the Flying Cloud Sanitary Landfill, and requests that an order be entered dismissing the contested case proceeding as moot, pursuant to Minn. R. 1400.5500 (K)."

Jerri burst into tears. She cried a full ten minutes before pulling herself together to go back home. By now, she abandoned her grocery list and shopping expedition. When Jerri stopped for the stop sign at Homeward Hills and Silverwood, the entrance to my former subdivision, tears flooded her eyes again. She couldn't continue for several minutes. Finally, she made it the last few blocks to her house on Phaeton Place, and telephoned her husband. The media called minutes later, wanting to come out for interviews.

At 1237 Stonegate Road, Hummelstown, Pennsylvania, the telephone rang, too. There was no one home. The answering machine took the message. The phone rang again, and then again, all morning and afternoon. The tape on the recorder filled up.

News roared faster than a wildfire through Eden Prairie and the State of Minnesota. Students from kindergarten through high school cheered after the news was announced on school intercoms. The local bake shop made a cake with the "Just Say No To BFI" logo, and sent it on to city hall. Champagne was brought in. And a noon press conference kicked off a party at city hall. In the neighborhoods around the dump, mothers spilled out into the streets to share the news and drink champagne, too. The press swarmed over town to report citizens' reactions.

BFI chose election primary day to disclose its permit withdrawal, hoping political events would top the news. One last time, they miscalculated. Beginning at noon, the Flying Cloud issue was the first news item on every state broadcast. The President of Woodlake Sanitary Service, Inc. calmly cited "damage to credibility" as the reason for his company's decision.

At 3:30 p.m., I finished my day as a teacher's aide at Hershey Elementary School and drove home. On the way, I remembered I had a 4:00 o'clock orthodontist appointment for Pierre, so I speeded up. It'll be tight, I thought, so I decided to pull into the driveway, keep the motor running, and beep the horn for Pierre to come out.

I honked and Pierre popped out.

"You have an orthodontist appointment, Pierre. Hop in the car!" I said.

"OK, but come in the house first, Mom," yelled Pierre.

"There's no time. You have an orthodontist appointment in a few minutes. Get in the car!"

He disobeyed. He kept jumping up and down, telling me to come in the house. I just got more angry. Pierre was probably playing one of his dumb jokes, I guessed.

"Get in the car right now, Pierre," I shouted. "We have to go."

"The dump is shut," said Pierre. "Listen, it's on the answering machine."

"We don't have time for your jokes; get in the car," I ordered.

By now I jumped out of the car to yell. Pierre then proceeded to pull me by the arm, some people can be so recalcitrant. That boy was persistent. He turned on the answering machine for me to hear. "It's 10:20 a.m., September 11, 1990," announced Sidney slowly and softly. "BFI threw in the rag. It's over. BFI withdrew its permit application." Grant came in next, "BFI gave up; they threw in the towel," he said, hurriedly, in his rough voice. Many more messages followed. I was finally convinced Pierre was telling the truth.

My three sons stood in bewildered silence as they watched me dissolve into a crying, hysterical mother, slamming my fists on the kitchen counters, saying "I can't believe it."(Oh, if the BFI boys could have seen me behaving like the emotional creature they accused me of being!) Paul said, "Mom, we knew you'd be excited, but this is ridiculous!"

I can't even remember how long I carried on, until I called Ivan to give him the news. Then I telephoned the orthodontist to apologize for our missed appointment. I tried explaining to the receptionist why. The fact we shut the Eden Prairie dump probably made little sense to the woman, but frankly, I didn't care.

Friends and the press rang my phone off the hook for the remainder of the day. We rejoiced together over our good fortune, and I told the *Eden Prairie News,* "Eden Prairie has just made environmental history. It's going to send a message around the world."

As the day wore on, I realized I hadn't heard from Jerri Coller. The telephone lines to her house must be busy, I thought, but I wanted to hear her voice, her reaction to the permit withdrawal, and congratulate her on the victory. About 9:00 p.m. I got through.

Since both of us had cried ourselves to exhaustion, we were pretty mellow and very emotionally drained at this point. We exchanged stories of our reaction to the news, and found we behaved similarly, chiding bearers of good tidings for playing a nasty joke, until they provided solid evidence of their honesty. We nearly dehydrated from all our weeping following the announcement.

Even though the initial shock of the news had worn off, we still questioned whether the crusade was really finished. Perhaps today was all just a dream. Tomorrow, we'd wake up to face the fight once again. We kept repeating, "I can't believe it's over."

Our conversation didn't last long. We were both very tired.

"Jerri," I asked, "if you had it to do over, would you fight the dump?"

"Absolutely! I'd do anything to protect my children," she answered. "They're the most precious of all I have."

I fully agreed with her.

In the name of our children and all children, it was worth our years of sacrifice. When we put our heads down to sleep that night, a feeling of peace overwhelmed us. We knew we hadn't walked this path in vain.

EDEN PRAIRIE SAILOR

EDEN PRAIRIE NEWS

14 | EPILOGUE

Bill and Dorrie Schmidt were relocated at BFI's expense, and the company bought the drive-in theater.

As a result of our victory, many other cases involving landfill sitings and expansions were decided in favor of the people and environment. Both Jerri and I mentored many protest group leaders. In the six months following the landfill shutdown in Eden Prairie, BFI's stock on Wall Street plunged from 42 to down to 16. Both the *Wall Street Journal* and *Forbes Magazine* did feature stories on the great fall of Browning Ferris Industries. CEO Ruckelshaus revealed that the company took a pretax charge of 36.5 million dollars in fiscal 1990's fourth quarter, for the Flying Cloud Landfill loss. *Good Housekeeping*, *Nation's Cities Weekly*, and other national publications published related articles.

In Minnesota, Sidney Pauly helped institute a "bad boy law," which prohibits a company with a criminal record from doing business in the state. Also, the state landfill siting process was stopped and a state mandatory recycling program established. The state reduced it's projected garbage output by 80%! Minnesota is now regarded as national leader in solid waste management.

Both Jerri Coller and I have received many awards for our leadership role in this case. Jerri continues to run the Homeward Hills Homeowners Association and monitor the landfill clean-up now in progress.

The current Midwest management at BFI is more responsive to people and their environment in Eden Prairie. The company now devotes considerable resources to community recycling programs in Minnesota and around the country.

Unfortunately, our environmental quest isn't an isolated story. Too many other people across the United States and around the world are also waking up to discover pollution in their backyards. Like the canaries in yesterday's mines, children are first to become ill and die when toxins are present in the environment. Their tolerance is lower than a healthy adult male for whom the standards are set. Locate the dumps, chemical factories, and nuclear facilities, and you'll find mothers caring for sick and dying children. It's that simple and commonplace. Ask Lois Gibbs. Regrettably, she says business is brisk at the Citizen's Clearinghouse for Hazardous Waste, the organization she founded to help people act against pollution and polluters in their neighborhoods.

We were lucky in Eden Prairie, because we averted disaster before the population became seriously sick. We won because we had the will, health, education, money, and time to outlast our opponents; but it was brutally hard work.

I wonder how people defend themselves if they must struggle daily to nurse an unhealthy baby, or need two incomes to put food on the table? How well can they decipher the scientific and legal jargon thrown at them? And how did that Hispanic woman manage, the one who lives in government subsidized housing on top of a hazardous landfill with her neighbors who can't speak English? In most cases, poorly. Perhaps this book may represent their voices, too; their cries of help to protect their children, and their rage at being made victims of pollution for the greed of industry and expediency of politics.

Hildegard of Bingen, a mystic who lived 800 years ago in Germany, warned:

> *The earth should not be injured.*
> *The earth should not be destroyed.*
> *As often as the elements,*

the elements of the world are violated
by ill treatment,
so God will cleanse them thru the sufferings,
thru the hardships of mankind.

It's already happening. We, the children of Mother Earth are becoming sick, because our mother is no longer healthy. Her water, air, and land which nourish us are polluted.

Worldwatch Institute, a government-funded, scientific research group, predicts the collapse of our civilization in 2040, based on the rate at which we strip the land of natural resources, contaminate the air, land, and soil, and overpopulate the planet. Very sadly, we're allowing industry and government to march us to Armageddon.

It disturbs me that in the United States of America, the country which promotes itself as the democratic leader of the world, it took eight and a half years, the same time it took George Washington to win the Revolutionary War, to accomplish what the Romans figured out 2000 years ago, Indians understood 1,000 years ago, Hildegard of Bingen knew 800 years ago, and Rachael Carson reiterated 30 years ago. Why aren't we clever enough to reread our world history books and learn from humankind's mistakes?

Instead, we/ve created giant monuments as testimony to our excessive consumption: toxic, leaking, garbage dumps. Today, the largest man-made monument on earth is the Fresh Kills Landfill in New York, unlovingly referred to as Mount Trashmore by local townspeople. Predictably, it's contaminating water, land, and air.

As this book goes to press, a thorough scientific study has just been completed by the New York Department of Health and researchers at Yale University. This study concludes that it's dangerous to live near toxic waste sites! The rate of birth defects, central nervous system defects, musculoskeletal system and skin problems were significantly higher in people living adjacent to toxic landfills than in the general population. How clever we are to confirm what others before us have known for centuries. And how terribly pleased I am that people followed their gut instincts, averting major health problems in Eden Prairie by standing firm in their defiance to reject expansion of a toxic, leaking landfill that's now a superfund site.

291

With Mother Earth in peril, we must seek opportunities in each community to safeguard the environment for all families, now and in the future. Time isn't on our side. Degradation of our planet is occurring rapidly. The only hope we have of rescuing it, is our determination to change the course of events by standing together with others and making demands of our leaders. It's my trickle-up theory of environmental pollution control, and it works. We, the ordinary folk of Eden Prairie, followed our vision and succeeded, despite overwhelming odds against us. You can do the same in your neighborhoods. Together, we can create a healthier future for all children. The future of civilization depends on our success.

EDEN PRAIRIE SAILOR (MINNESOTA SUN PUBLICATIONS)

Photo Credits

Page 17 aerial photographs

MARKHURD. *Flying Cloud Landfill area*. Minneapolis, Minnesota: Hennepin County, 1945.

MARKHURD. *Flying Cloud Landfill area*. Minneapolis, Minnesota: Hennepin County, 1989.

Chapter 1, page 21

Varlamoff, Susan. *Danger sign on perimeter fence at Flying Cloud Landfill*. Eden Prairie, Minnesota: Personal photograph.

Chapter 2, page 27

Greener, Stormi. *Susan Varlamoff with her sons Neil, left, and Paul*. Mpls.-St. Paul, Minnesota: Star Tribune, Aug. 1987.

Chapter 3, page 45

Eden Prairie News. *'Ring around the dump' protest*. Eden Prairie, Minnesota: Eden Prairie News, June 1986.

Chapter 4, page 65

Greener, Stormi. *Members of the Minnesota Pollution Control Agency conferring at a public meeting about Flying Cloud Landfill*. Mpls.-St. Paul, Minnesota: Star Tribune, June 27, 1990.

Chapter 5, page 81

Mueller, John. *Mock funeral rally at the Minnesota state capitol*. Eden Prairie, Minnesota: Eden Prairie News, April 1989.

Chapter 6, page 105

Hager, Art. *Protesters use a bus to block the entrance to BFI's Flying Cloud Landfill in protest of its proposed expansion*. Mpls.-St. Paul, Minnesota: Star Tribune, Feb. 16, 1988.

Chapter 7, page 129

Varlamoff, Susan. *Mural created by children depicting recycling vs. the landfill disposal of wastes*. Eden Prairie, Minnesota: Personal photograph.

Chapter 8, page 151
Office of Minnesota State Rep. Sidney Pauly. *Susan Varlamoff and State Rep. Sidney Pauly in the legislative chamber* . St. Paul, Minnesota: Courtesy Minnesota State Legislature.

Chapter 9, page 167
Varlamoff, Susan. *The Varlamoff home.* Eden Prairie, Minnesota: Personal photograph.

Chapter 10, page 197
Dick, Betsy. *Greg Swaim, 4, is held up to the podium by his father, Bill, and his mother, Bonnie, so he could testify at a public hearing session held at Eden Prairie High School.* Eden Prairie, Minnesota: Photo reprinted from Eden Prairie Sailor newspaper (Minnesota Sun Publications), Dec. 1988.

Chapter 11, page 223
Mueller, John. *Jerri Coller prepares for public meetings on the Flying Cloud Landfill.* Eden Prairie, Minnesota: Eden Prairie News, Nov. 1, 1989.

Chapter 12, page 249
Mueller, John. *Dick Coller leading "close the dump" chant at Pax Christi Church.* Eden Prairie, Minnesota: Eden Prairie News, Nov. 22, 1989.

Chapter 13, page 273
Weber, Mark. *Protesters parading along perimeter of Flying Cloud Landfill.* Eden Prairie, Minnesota: Eden Prairie News, June 1986.

Chapter 14, page 289
Cullen, Linda. *Protester Richelle Reid at the state capitol.* Eden Prairie, Minnesota: Photo reprinted from Eden Prairie Sailor newspaper (Minnesota Sun Publications), Apr. 1989.
Weber, Mark. *Protester Andy Wilson.* Eden Prairie, Minnesota: Eden Prairie News, June 1986.
June 1986.

Chapter 14, page 293
Lassig, Craig. *Renee Coller at memorial vigil.* Eden Prairie, Minnesota: Photo reprinted from Eden Prairie Sailor newspaper (Minnesota Sun Publications), Apr. 1988.

SOURCES

Chapter 2

■ Anderson, Helen Holden, *Eden Prairie,The First 100 Years,* Minnesota, Viking, 1979.

■ Carson, Rachael, *Silent Spring*, Boston, Houghton Mifflin, 1987.

■ Mengel, Joseph, Technical Memo on Eden Prairie Landfill Site, 1988.

Chapter 3

■ Analytical Summary of the Flying Cloud Landfill, Pace Laboratory, October 13, 1986.

■ Chiras, Daniel D., *Environmental Science*, The Benjamin /Cummings Publishing Co., Inc., California, 1988.

■ Hammer, Mark J., Viessman Jr., Warren, *Water Supply and Pollution Control*, Harper and Row, New York, 1985.

■ Jacobs, Paul, "Study Finds Toxic Gases In Ordinary Trash Dumps," *Los Angeles Times*, November 25, 1986.

Chapter 8

■ Damen, Al, "Justices to Consider Limiting Punitive Damages," *The Washington Post*, December 6, 1988.

Chapter 10

■ Congress of the United States Office of Technology Assessment, "Are We Cleaning Up?" Special Report, OTA-ITE-363, Washington, D.C., June, 1988.

■ Lehman and Associates, Inc., "Response and Supporting Statement of the City of Eden Prairie and The Homeward Hills Homeowners Association," April 18, 1989.

Chapter 11

■ Association of Bay Area Governments, "Air Emissions Associated With Publically Owned Treatment Works in Santa Clara Valley," May, 1985.

■ Flower, Franklin B.; Leone, Ida A.; Gilman, Edward F.; Arthur, John J.; "Vegetation Kills in Landfill Environs," 1977.

■ Jacobs, Paul, "Study Finds Toxic Gases In Ordinary Trash Dumps," *Los Angeles Times*, November 26, 1986.

■ Ozonoff, David, "Silresim Area Health Study," November 22, 1983.

■ Ryan, Christopher M.; Morrow, Lisa; Hodgson, Michael; "Cacosmia and Neurobehavioral Dysfunction Associated With Occupational Exposure to Mixtures of Organic Solvents," *American Journal of Psychiatry*, 145:11, November, 1988.

Chapter 13

■ Hertzman, Clyde; Hayes, Mike; Singer, Joel; Highland, Joseph; "Upper Ottawa Street Landfill Site Health Study," *Environmental Health Perspectives*, Volume 75, pages 173-1951, 1987.

Chapter 14

■ Gesschwind, Sandra; Stolwijk, Jan; Bracken, Michael; Fitzgerald, Ed; Stark, Alice; Olsen, Carolyn; Melius, James; "Risk of Congenital Malformations Associated With Proximity to Hazardous Waste Sites," *American Journal of Epidemiology*, Volume 135, No. 11, Pages 1197-1207, August, 1992.

ALSO FROM *ST. JOHN'S PUBLISHING* . . .

● ● ●

Parenting a Business, by Donna L. Montgomery, looks at business relationships from a parenting standpoint.

Surviving Motherhood, by Donna L. Montgomery. A look at family relationships written by a mother of eight who is a survivor of motherhood herself.

Kids+ Modeling= Money, by Donna L. Montgomery, is all you need to help your child begin a rewarding and prosperous modeling career. Discover the secrets of modeling success.

● ● ●

ST. JOHN'S PUBLISHING
6824 OAKLAWN AVENUE
EDINA, MN 55435

Please send me _____ copy (copies) of **Parenting a Business** (ISBN 0-938577-04-2). I am enclosing $14.95 and $1.50 for shipping for each copy.

Please send me _____ copy (copies) of **Surviving Motherhood**, (ISBN 0-938577-00-X). I am enclosing $6.95 and $1.50 for shipping for each copy.

Please send me _____ copy (copies) of **Kids+ Modeling= Money,** (ISBN 0-13-515172-4). I am enclosing $9.95 (hardcover) and $1.50 for shipping for each copy.

NAME _____

ADDRESS _____

_____

_____

_____